DAIRY CATTLE JUDGING TECHNIQUES

FOURTH EDITION

GEORGE W. TRIMBERGER

Professor Emeritus of Animal Science
New York State College of Agriculture & Life Sciences
Cornell University

WILLIAM M. ETGEN

Professor of Dairy Science
College of Agriculture & Life Sciences
Virginia Polytechnic Institute and State University

DAVID M. GALTON

Associate Professor of Animal Science
New York State College of Agriculture & Life Sciences
Cornell University

PRENTICE-HALL, INC., Englewood Cliffs, New Jersey 07632

Library of Congress Cataloging-in-Publication Data

TRIMBERGER, GEORGE W.
    Dairy cattle judging techniques.

    Includes index.
    1. Dairy cattle—Judging.   I. Etgen, William M.
II. Galton, David M.   III. Title.
SF215.2.T74 1987   636.2'0811   86-9336
ISBN   0-13-196007-5

Editorial/production supervision: *Erica Orloff*
Cover design: *Lundgren Graphics, Ltd.*
Manufacturing buyer: *Barbara Kittle*
Page layout: *Irene Poth*

Printed in the United States of America

10 9 8 7 6 5 4 3 2 1

ISBN 0-13-196007-5 025

Prentice-Hall International (UK) Limited, *London*
Prentice-Hall of Australia Pty. Limited, *Sydney*
Prentice-Hall Canada Inc., *Toronto*
Prentice-Hall Hispanoamericana, S.A., *Mexico*
Prentice-Hall of India Private Limited, *New Delhi*
Prentice-Hall of Japan, Inc., *Tokyo*
Prentice-Hall of Southeast Asia Pte. Ltd., *Singapore*
Editora Prentice-Hall do Brasil, Ltda., *Rio de Janeiro*

ii

CONTENTS

THE COVER

The threesome pictured on the cover represent the epitome in dairy cattle breeding for production and type. All three of the cows have extraordinary production, have reached the highest score of 97 for the breed in herd classification and have been voted All-Time All Americans, with each one of them having won at least three All-American Awards. Never before in the history of dairy cattle breeding has it been possible to make the above statements. They are the best of all time. Some specific information follows:

Shadowcliff R A Gina   All-American 2-Year-Old All-Time
All-American 2-Year-Old   All-American 4-Year-Old (cover)

1y-10m	285d	16,950	lb milk	3.9%	668 lb fat
2y-9m	310d	23,070	lb milk	3.6%	858 lb fat
3y-9m	339d	23,343	lb milk	3.4%	788 lb fat

Shadowcliff R A Gina, All-American 3-Year-Old, All-Time
All-American 3-Year-Old, Reserve All-Time All-American 4-Year
Old (cover)

4y-9m	327d	18,460 lb milk	3.9%	719 lb fat
5y-9m	365d	27,320 lb milk	3.5%	958 lb fat

(Courtesy Bower Farms, Lagrangeville, NY and Catskill Registered
Holsteins, Walton, NY)

Brookview Tony Charity, All-American 4-Year-Old, All-Canadian
4-Year-Old, All-Time All-American 4-Year-Old, All-Canadian
5-Year-Old, Twice All-American Aged Cow (unanimous), Twice
All-Canadian Mature Cow

2y-4m	365d	19,698 lb milk	4.0%	785 lb fat
3y-7m	329d	21,789 lb milk	3.9%	884 lb fat
4y-11m	343d	37,340 lb milk	3.5%	1310 lb fat
5y-11m	365d	39,015 lb milk	3.6%	1422 lb fat

(Courtesy Hanover Hill Holsteins, Port Perry, Ontario, and Millerton, NY)

Northcraft Ella Elevation, All-American 3-Year-Old, Reserve All-Time
All-American 3-Year- Old, All-American Aged Cow, 3 successive years,
All-Canadian Mature Cow, All-Time All-American Aged Cow

2y-9m	364d	32,803 lb milk	3.4%	1101 lb fat
4y-7m	328d	35,515 lb milk	4.0%	1411 lb fat
5y-9m	365d	44,143 lb milk	3.8%	1698 lb fat
7y-7m	365d	48,731 lb milk	4.2%	2028 lb fat

Number 1 cow CTPI at +1088 Cow Index +2,185 lb milk +.14% +106
lb fat +$318. (Courtesy Woodbine Farms, Airville, Pa and Romandale
Farms Ltd., Unionville, Ontario)

PREFACE

Dairy Cattle Judging Techniques, 4th edition, is changed in many respects to make it current. A new author, David M. Galton, who recently received the Professor of Merit Award from the Cornell College of Agriculture and Life Sciences, has been added. A new chapter added relates entirely to 4-H and FFA youth judging, including a reason class by a 4-H participant who was high individual for all breeds in the National Contest. This follows the pattern that has always been prevalent for this book—to emphasize accomplishments based on a proven record.

Starting with the cover page and continuing throughout the book, emphasis has been placed on outstanding production, which is basic, and on functional type traits, which are so important for wearability. It is only from this combination that dairymen can get sustained and efficient production for a successful and profitable operation.

The discussion is always on the usefulness of functional type traits in breeding programs; the association of these components with high production and longevity; and the association of type with production for cattle sales. Never before have production and type been combined in such a forceful way to achieve unprecedented results. *Dairy Cattle Judging Techniques* has always been noted for containing many pictures to illustrate dairy cattle evaluation and judging. To further enhance on this, some 40 new pictures have been added in this revised edition. Thus, it continues to be the best-illustrated book in the field.

The ideal, modern type for each breed is clearly defined, and a full discussion is given to each part of conformation, always stressing the intimate relationship between ideal functional conformation and high lifetime productivity.

From their extensive experiences at all levels, the three authors involve the practical approach. The subject matter is so grouped that the book is useful for dairy cattle breeders; commercial dairy farmers; professional judges; college, 4-H, and FFA judges; and anyone else with a special interest in dairy cattle. Everything is complete for precise type evaluation, selection within the herd, show-ring judging, fitting and showing, show-ring procedures, placings and reasons in youth (4-H and FFA) and college student judging contests. Type classification procedures are described in detail.

This new edition of *Dairy Cattle Judging Techniques* could not have been written without the help and cooperation of many breeders and friends and the suggestions of fellow dairy judging team coaches. The authors gratefully acknowledge their assistance and stimulating influence. The breed associations have contributed generously from their supply of material. Particular acknowledgement is made of the skill and cooperation of dairy cattle photographers. It was through their excellent work that many outstanding photographs from breeders of purebred cattle were available to supplement the text.

The authors greatly appreciate the suggestions of colleagues at Cornell University and Virginia Tech, as well as the wholehearted cooperation of many breeders who provided pictures of cattle to illustrate the judging techniques in this book. Most of all, we appreciate the privilege of working with the fine young people who, as judging team members, have provided us with many of our most satisfying experiences. We thank the breeders for their extra efforts in providing cattle for judging classes and thereby cooperating with this constructive, educational work.

George W. Trimberger
William M. Etgen
David M. Galton

1
A PRACTICAL PHILOSOPHY
OF JUDGING

In order to be a good breeder and feeder and to cope with management problems, a dairy farmer must know how to judge and observe his or her cattle. If the dairy farmer cannot evaluate his or her cattle, individually and as a group, he or she lacks one of the basic requirements for a successful herd manager. Such knowledge is of practical use in choosing foundation cattle, in buying and selling, in culling to improve the herd, in selecting herd sires, and in handling cows to keep them at peak production.

A good judge knows what characteristics are required for a long, useful life of high production and a low incidence of feet and leg problems and udder disturbances. Honnette, Vinson, and White[1] listed several functional type traits that are significantly correlated with longevity and lifetime milk production in U.S. Holsteins: strong udder support; high, wide rear udder; ideal teat size and placement; medium strength front end; and narrow front end. They also listed several traits associated with significantly less than average longevity and lifetime production. These were the following: coarse front end, crampiness, broken udder support, teats too large, and loose rear udder.

Dairy cattle judging has been instrumental in arousing in many young people an interest in good dairy cattle.

[1]J. E. Honnette, W. E. Vinson, and J. M. White, "Relationships Between Descriptive Type Traits, Milk Production, Calving Interval and Herdlife in Holsteins," Proc. 70th Ann. Meet. Amer. Soc. Anim. Soc., 1978, p. 236.

Figure 1A George W. Trimberger (center) teaching a selected group of students the important points for precision judging. This approach is necessary for uniform and outstanding accomplishments. Half of his 24 Cornell teams placed among the top three in national competition, and there were seven winners of the National Intercollegiate Contest. Also included was a triple winner with firsts at the Eastern States Exposition, Springfield, Massachusetts; the All-American, Harrisburg, Pennsylvania; and the sweepstakes at the National Intercollegiate, Columbus, Ohio. (Courtesy Don Bay, Macedon, NY)

The technique of type evaluation is based on careful observation, thoughtful evaluation, and intelligent decision making founded on organized thinking. Another part of judging activities is delivering oral or written reasons. Good reasons, which are the justification of the decision, are accurate, thorough, and well organized and are delivered in a confident manner. Learning to judge dairy cattle and give reasons helps develop confidence in one's ability to make and defend good decisions—skills needed in almost every aspect of life. Training students and others to select dairy cattle for high production with functional type ensures a more efficient dairy operation and benefits both producers and consumers.

THE TECHNIQUE OF OBSERVING AND COMPARING WITH THE IDEAL

To be successful in evaluating and placing the individual animals in a class, one must have a thorough knowledge of the desirable conformation of all parts of the animal. This serves as an ideal or standard with which comparisons are made. The technique of judging is a combination of science and art, and it is dependent on the precision with which selections conform to a definite standard or pattern.

Once the judge develops the ability to see an animal quickly and accurately by recognizing all of the good and bad points, the judge must then learn the technique of making a good, sound judgment based on practical reasoning. Because there are so many variations among individuals and classes, a set

Figure 1B William M. Etgen (standing right), M. L. McGilliard (kneeling center); and the eight students who comprised the Virginia Polytechnic Institute dairy judging teams which placed first in all four contests they entered in one year (Eastern States, Pennsylvania All-American, Mid-South, and the National Contest). Richard Morris (standing left) was high individual in the National Contest.

pattern for placing animals in the various classes is not practicable. The final decision involves practical reasoning and good judgment in comparing animals with the ideal or standard animal. Thus, judging provides fine experience and excellent training at all levels of participation. Judging is a stimulating game in which one observes the appearance and characteristics of a large number of animals and then ranks them comparatively on merit or desirability of type.

A good judge must be able to form a clear picture of a large number of individuals and retain the observations clearly, so he or she can make comparisons and rank a class largely from memory. To judge intelligently, the judge must have an intuitive knowledge of livestock.

The best judges usually have considerable experience with dairy cattle. This provides for unlimited observations of the merits and demerits of individual animals. To observe how certain characteristics change during the normal lifetime of an animal is both interesting and educational. This is demonstrated by Figures 2 through 5, which show the great Jane of Vernon at 3, 4, 11, and 15 years of age. This outstanding cow has meant more to the future of the Brown

Figure 1C David M. Galton (front right) and the Cornell University team that placed third at the National Contest. David King (center) was second high individual overall and high individual in Holstein. David Chlus (far left) was high individual in Jerseys. (Courtesy David M. Galton, Cornell Univ. Ithaca, N.Y.)

Figure 2 Jane of Vernon, Wisconsin Grand Champion at the age of 3 years. Her outstanding dairy quality, the smoothness of her body conformation, and the unusually well-shaped and well-attached udder at this young age show that she is destined to become a great cow. (Courtesy Brown Swiss Breeders Association, Beloit, WI)

Figure 3 Jane of Vernon at the age of 4 years when she won the Grand Champion award at the Dairy Cattle Congress. She repeated this until she had won for 5 consecutive years. At the age of 4 years she also set the world's record with 23,569 lb of milk and 1075.6 lb of fat. (Courtesy Brown Swiss Breeders Association, Beloit, WI, and reprinted with permission from John Wiley & Sons, Inc., New York, NY)

Figure 4 Jane of Vernon at the age of 11 years. Her marvelous type gives her the bloom and smoothness of a younger cow. At this age, her desirable dairy character, smoothness and blending of parts, the depth and neatness of barrel, the nearly perfect udder, and the strength of bone in her ideal legs do not show any wear from her world-record performance as a young cow. Other high records that followed gave her two records of over 1000 lb of fat each, with well over 20,000 lb of milk. (Courtesy Brown Swiss Breeders Association, Beloit, WI)

Figure 5 Jane of Vernon at the advanced age of 15 years still displays unbelievably good type. Her head, neck, shoulders, body, topline, and legs show desirable signs of advanced age. The dip in her loin and pelvic region results from age. However, her udder displays faultless attachments, unusually good suspensory ligament resulting in levelness of floor, and an ideal teat arrangement. (Courtesy Brown Swiss Breeders Association, Beloit, WI)

Swiss breed than any other individual dairy cow has meant to its own breed. Her greatness was in her excellent type, her producing ability, her transmitting qualities, her longevity. In her long lifetime she effected a great improvement in the breed through her many descendants.

A good judge must know the general type that prevails in the cattle population. The judge should observe a large number of cattle in order to learn what constitutes a normal population. Once a judge is able to recognize and decide whether a particular individual is superior or inferior, he or she can decide whether the individual ranks in the upper 25, 10, or 5 percent of the population or whether she is average or even inferior when compared with the entire cattle population.

PRACTICAL ADVANTAGES OF GOOD TYPE

A good dairy cattle judge must have a practical viewpoint and base selections for individual cattle on the purpose for which they are used. Emphasis on utility is important in the selection and improvement of dairy cattle. In using the analytical approach to evaluate the good and bad points, a good judge notes those that will improve with time and discriminates against those that will become less desirable with age.

If the proper emphasis in the selection of dairy cattle is placed on functional type, the kind of a cow that can stand up under high production year after year should result. Published data show that for all breeds the average dairy cow remains in the herd for fewer than 5 years after she reaches productive life, and the cumulative dry periods during this time usually add up to more than 1 year.

The practical advantage of the herd owner and dairy cattle breeder for emphasis on both milk yield and classification score in a selection program was clearly demonstrated in research by Dr. R. H. Kliewer of the Holstein-Friesian Association of America in cooperation with research by staff members at North Carolina State University. The difference shows up in decreased and increased herd life. When milk yield alone is used in selection (top 10 percent), the length of herd life is decreased by 153 days in five generations and by 306 days in ten generations. In contrast, when selections are based on both milk yield and type classification score there is an increase of 255 days in length of herd life in the fifth generation and 510 days above the average in the tenth generation. The results are presented in detail on the following page.

Type is best appreciated when differences from old age and wear start to appear. Many cows are fine in all respects up to 5 or 6 years of age and then suddenly they begin to decline. At this time, type (udder conformation, foot and leg structure, muscle tone, constitution or general strength, and many other characteristics of conformation) becomes increasingly important for longevity or sustained lifetime production.

Many of the body characteristics that are stressed in considerations of de-

Selection Trait(s)	Generation			
	Present	1st	5th	10th
Milk Yield Alone				
Milk Yield (pounds)	15,000	15,418	17,090	19,180
Classification Score (points)	79.8	79.2	76.8	73.8
Herd life (days)	6.0 (yr, mo)	− 51	− 153	− 306
Milk Yield and Final Classification Score Combined				
Milk Yield (pounds)	15,000	15,320	16,600	18,200
Classification Score (points)	79.8	80.8	84.7	89.9
Herd Life (days)	6.0 (yr, mo)	+ 51	+ 255	+ 510

sirable type play an important part in high production. For example, a good dairy cow has little excess flesh on her frame. If she is to move about freely from feed bunk to milking parlor to resting area or to and from pasture, she must have feet and legs of proper structure and soundness. This is especially important in confined housing systems which are becoming increasingly popular in the United States. Here dairy cows spend considerable time walking and standing on a hard surface such as concrete. The proper kind of udder—that is, one with strong udder support; high, wide rear udder; and ideal or nearly ideal teat size and placement—is necessary for sustained high production and resistance to injury and disease to which all cows are exposed from time to time.

Type is more important today than ever before because disease control measures permit special emphasis on having the best kind of cow for high production during a long lifetime. The development of practical embryo transfer techniques permits continued production of many offspring from superior cows. Good type should be combined with breeding for high production, proper disease resistance, and recognized management procedures to permit full and efficient use of animals having outstanding conformation. When good type appears with these other qualities, eye appeal helps to sell surplus females. That the owner and his or her employees get more satisfaction and enjoyment from cattle of good type is often reflected in better care and higher economic returns.

Dairy cattle breeders in foreign countries have been recognizing and emphasizing type for many years. That foreign breeders will continue this emphasis is indicated by the importance attached to functional type when cattle are exported from the United States. The higher prices prevailing in the United States for dairy cattle of good type indicate a general acceptance of the importance of this factor by dairy farmers. Dairy cattle developed in the United States are considered to be the best in the world.

In a recent study by Ruff and others at Virginia Polytechnic Institute, the direct value of type was measured in terms of increased returns from cattle sales for superior type purebred Holsteins.[2] A summary of the results of the

[2]N. J. Ruff, and others, "Factors Affecting Within-Sale Price Differences in Holsteins," Proc. 75th Ann. Meet. Amer. Dairy Sci. Assoc., 1980, p. 104.

study of 11,220 animals sold in 167 sales held during the years 1973–1977 are presented.

From these data it can be seen that the factors having the largest effects on sale prices of heifers were the heifer's dam's type score and fat yield and her sire's PD for milk, test, and type. The total percent of the sale price determined by type factors for heifers was 44.9 as compared to 55.1 for production factors. In the case of cows, type factors contributed 44.1 and production factors 56.0 percent of the relative importance to sale price. In summary, type factors contributed nearly as much as production factors to final sale price.

Percent Contribution to Sale Price

	% for Heifers	% for Cows	MGD[a]	% for Heifers	% for Cows
PD[b]milk	13.4	7.3	Milk yield	3.7	—
PD fat %	10.6	6.1	Fat yield	8.0	—
PD type	14.6	9.0	Type score	6.1	—
Dam			Cow		
Milk yield	1.7	13.0	Milk yield	—	16.4
Fat yield	11.3	8.7	Fat yield	—	0.1
Type score	15.1	5.3	Type score	—	23.0
MGS[c]					
PD milk	3.0	1.1	Total % production	55.1	56.0
PD fat %	3.4	3.3	Total % type	44.9	44.1
PD type	9.0	6.8	Average sale price	$1835	$2189

[a]MGD = maternal granddam

[b]PD = predicted difference

[c]MGS = maternal grandsire

PRODUCTION AND FUNCTIONAL TYPE

The statement on production and type at the start of Chapter 25 has received so much attention and appreciation that the authors decided to express the same idea in Chapter 1—A Practical Philosophy of Judging as follows:

> Production is basic and must receive first consideration, however, functional type must receive proper attention in order to capitalize on the high production bred into the modern dairy cow. Good type with poor production is relatively useless and a high producer with unacceptable deficiencies in type resulting in poor wearability with short lifetime production is inefficient and not economical.

Type should not be a substitute for but rather a valuable addition to high production. All dairy farmers recognize the importance of utility or functional type traits such as dairy character; udder attachments, especially the suspensory ligament; and feet and legs. The purebred breeder adds a few to these, for example, nearly level rumps, attractive heads, and several traits which are more or less show-ring-type traits. The dairy farmer also keeps the practical usefulness of a type trait in mind. For example, rump specifications were changed from one perfectly level to a conformation in which the pins are slightly lower than the hips. This has an advantage for ease of calving and genital health because it is inducive to better drainage after parturition and improved sanitation of the genital system throughout the year.

QUALITIES OF A GOOD JUDGE

Desirable characteristics for success in judging dairy cattle are:

1. "Livestock-mindedness" and a desire to know dairy cattle thoroughly.
2. A clear knowledge of the ideal or standard type, and an ability to recognize desirable and undesirable points of conformation.
3. Quick and accurate powers of observation.
4. Ability to form a mental image of many individual animals and to rank them by making comparisons.

Figure 5A Cows at Bob Staley's Auction Sale in California, where buyers were among the first to demonstrate in a positive way an appreciation for functional type in grade and commercial cattle. The buyers paid higher prices for cattle with the right kind of udders, good feet and legs, dairy quality with strength in overall conformation, and other characteristics. Photo by senior author while he was in California working for the Holstein Association on the revision of the type classification system. (Courtesy George W. Trimberger, Cornell Univ. Ithaca, N.Y.)

5. Reasoning power that takes into account practical considerations.
6. Ability to reach a definite decision based on sound judgment.
7. Extreme honesty and sincerity, in order to avoid bias or prejudice. Judges are selected on character, including courage and honesty.
8. Steady nerves and confidence in one's ability to make close independent decisions based entirely on the merits of the animals. Students in practice and in contests should always work independently. A good philosophy for all judging is to do the best work possible at the time and to have no regrets about the results or accomplishments.
9. Evaluate and rank the individual animal according to her appearance on the day of judging, regardless of her rank at a previous show.
10. Sound knowledge acquired through practice and experience, in order to give effective reasons for decisions.
11. A pleasant and even temperament. Good judges, however, do not fraternize with exhibitors or friends along the ringside.
12. Firmness to stand by and defend one's placing without offending or in any way implying that one's decisions are infallible.

2
SKELETAL STRUCTURE AND PARTS OF A DAIRY COW

The extensive use of descriptive terms is necessary to make judging educational and to explain to others the judgment involved in placing a class. The ability to explain the major and minor differences that justify the placings of various individuals is as important as the final decision itself. This is possible only if the judge knows the name of every part of a dairy cow. (Fig. 6).

A good judge of dairy cattle must be so expert that he or she can literally "see through" a cow. In this respect, the ability to visualize the skeletal structure or framework of a cow is a distinct advantage. For example, when one evaluates such characteristics as smoothness of shoulder, width, length, and levelness of the pelvic region, or strength of the pasterns and set of the hind legs, the importance of the condition observed is much easier to decide upon if one has a fundamental knowledge of the bone structure supporting and shaping these parts (Fig. 7). In the final analysis, a judge must have a clear conception of the function of the skeleton in order to know the gross structural form that gives shape and size to the dairy cow and bull. The frame determines durability, strength, and longevity.

TEETH INDICATE APPROXIMATE AGE

Occasionally, a judge will want to determine the approximate age of an individual by examining the teeth. This is possible if one knows when the permanent teeth replace the temporary or milk teeth. G. W. Pope describes this in USDA Farmers' Bulletin 1721.

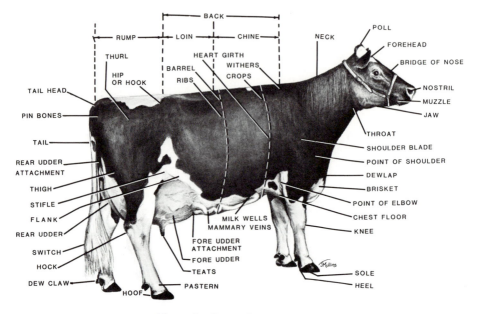

Figure 6   Parts of a dairy cow.

Figure 7   Skeleton of a dairy cow.

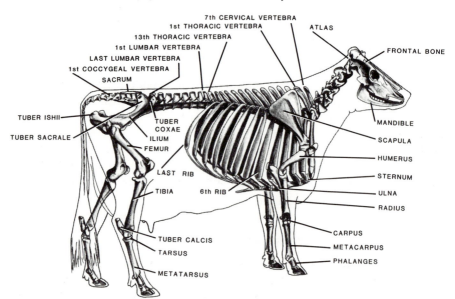

A mature cow has 32 teeth, eight of which are incisors located on the lower jaw. The dental pad, a thick layer of hard palate, takes the place of incisor teeth on the upper jaw. Two of the central incisors on the lower jaw are called pinchers; the next two, first intermediates; the third pair, second intermediates or laterals; and the outer pair, the corners.

The calf is born with two or more incisors of the first set, or temporary teeth, and the rest appear during the first month. The sequence in which the permanent teeth replace the first set (Fig. 8) when the cow is from 2 to 5 years of age is shown in Figures 9 through 12.

Beyond 5 years the approximate age can be guessed by the amount of wear indicated by the permanent teeth. Between 5 and 6 years the permanent pinchers begin to level. Wearing becomes more noticeable at 6 years when both pairs of intermediates become partially leveled and the corner pair begins to show wear. Noticeable wear of the pinchers is evident at 7 to 8 years; of the middle pairs, at 8 to 9 years; and of the corner teeth, at 10 years. As the cow advances in age from 6 to 12 years, the rounded contour of the teeth is gradually lost, and the line of the lower teeth becomes nearly straight by the twelfth year. Usually, at 12 years the teeth are distinctly separated, triangular in shape, and worn down to stubs (Fig. 13).

Figure 9   As the heifer approaches 2 years of age, the central pair of temporary incisor teeth (pinchers) is replaced by the permanent pinchers, which usually attain full development at 2 years.

Figure 8   All of the temporary or milk teeth have appeared in the calf at the age of 1 month.

Figure 11   The second intermediates, or laterals, are cut at 3½ years. They are on a level with the first intermediates and begin to wear at 4 years.

Figure 10   The permanent first intermediates, one on each side of the pinchers, are cut at about 2½ years and are usually fully developed at 3 years.

Figure 12   The corner teeth are replaced at about 4½ years, so at 5 years of age the cow has a full complement of incisors with the corners fully developed.

Figure 13   Constant wear has reduced the internal face of the incisors to stubs at 12 years of age. This condition becomes more marked with increasing age.

3
USE OF THE UNIFIED SCORE CARD FOR TYPE EVALUATION

The score card, periodically revised and brought up to date, has played a vital part in the development of the modern viewpoint on type. At one time each of the six dairy breeds had a separate and distinct score card, and no two were alike. Today, the points for evaluation of dairy type and body conformation are the same for Ayrshire, Brown Swiss, Guernsey, Holstein, Jersey, and Milking Shorthorn. This was accomplished through the work of a committee of the Pure-bred Dairy Cattle Association (PDCA).

The adoption of the unified score card was based on the fundamental concept that a good dairy animal has certain common characteristics that are more basic than her particular shape, color, size, and other breed characteristics. The committee acted wisely in devising a score card that uniformly considers the basic points for all breeds and, at the same time, takes into account the characteristics that account for the differences in the individuality and general appearance of each breed.

Adoption of the unified score card has also done much to correct the old problem of comparing in the show ring the distinctly different types that are found in different parts of the country. This difficulty has gradually disappeared, and now cattle from California and other western states, for instance, can compete with cattle from the East and the Midwest at a midwestern show. All breeders emphasize the same general points of body conformation and dairy refinement.

Score cards for dairy cows and dairy bulls have been developed and approved by the Purebred Dairy Cattle Association. These score cards are revised as needed to reflect changes in the conformation of the ideal and changes in the emphasis or value placed on the various type components.

DAIRY COW UNIFIED SCORE CARD[1]

General Appearance (35 Points)

Attractive individuality with femininity, vigor, stretch, scale, and harmonious blending of all parts with impressive style and carriage.

(5) *Breed Characteristics*—[see pages 17, 18]

(5) *Stature*—Height including moderate length in the leg bones with a long bone pattern throughout the body structure.

(5) *Front End*—Adequate constitution with strength and dairy refinement.

Shoulder Blades and Elbow—Set firmly and smoothly against the chest wall and withers to form a smooth union with the neck and body.

Chest—Deep and full with ample width between front legs.

(5) *Back*—Straight and strong; loin broad, strong, and nearly level and rump long, wide, and nearly level with pin bones slightly lower than hip bones.

Thurls—High and wide apart.

Tail Head—Set nearly level with topline and with tail head and tail free from coarseness.

(15) *Legs and Feet*—Bone flat and strong.

Front Legs—Straight, wide apart, and squarely placed.

Hind Legs—Nearly perpendicular from hock to pastern from a side view and straight from the rear view.

Hocks—Cleanly molded and free from coarseness and puffiness.

Pasterns—Short and strong with some flexibility.

Feet—Short and well rounded with deep heel and level sole.

Dairy Character (20 Points)

Angularity and general openness without weakness, freedom from coarseness, and evidence of milking ability with udder quality giving due regard to stage of lactation.

[1]This score card is reprinted with the permission of the Purebred Dairy Cattle Association.

Neck—Long, lean, and blending smoothly into shoulders; clean-cut throat, dewlap, and brisket.

Withers—Sharp with chine prominent.

Ribs—Wide apart; rib bones wide, flat, and long.

Thighs—Incurving to flat and wide apart from the rear view, providing ample room for the udder and its rear attachment.

Skin—Thin, loose, and pliable.

Body Capacity (10 Points)

Relatively large in proportion to size, age, and period of gestation of animal, providing ample capacity, strength, and vigor.

Chest—Large, deep, and wide floor with well-sprung fore ribs blending into shoulders; crops full.

Body—Strongly supported; long, deep, and wide; depth and spring of rib tending to increase toward the rear.

Flanks—Deep and refined.

Udder (35 Points)

Strongly attached; well balanced with adequate capacity possessing quality indicating heavy milk production for a long period of usefulness.

(6) *Fore Udder*—Strongly and smoothly attached; moderate length and uniform width from front to rear.

(8) *Rear Udder*—Strongly attached; high and wide with uniform width from top to bottom and slightly rounded to udder floor.

(11) *Udder Support*—Udder carried snugly above the hocks showing a strong suspensory ligament with clearly defined halving.

(5) *Teats*—Uniform size of medium length and diameter, cylindrical, squarely placed under each quarter, plumb, and well spaced from side and rear views.

(5) *Balance, Symmetry, and Quality*—Symmetrical with moderate length, width, and depth; no quartering on sides and level floor as viewed from the side; soft, pliable, and well collapsed after milking; quarters evenly balanced.

Because of the natural undeveloped udder in heifer calves and yearlings, less emphasis is placed on udder and more is placed on general appearance, dairy character, and body capacity. A slight to serious discrimination applies to over-developed, fatty udders in heifer calves and yearlings.

BREED CHARACTERISTICS

Except for differences in color, size, and head character, all breeds are judged on the same standards as outlined in the Unified Score Card. If an animal is registered by one of the dairy breed associations, no discrimination against color or color pattern is made.

Ayrshire

Ayrshires are strong and robust, show constitution and vigor, symmetry, style, and balance throughout, and are characterized by strongly attached, evenly balanced, well-shaped udders.

1. *Head:* Clean cut, proportionate to body; broad muzzle with large, open nostrils; strong jaw; large, bright eyes; broad and moderately dished forehead; bridge of nose straight; ears of medium size and alertly carried.
2. *Color:* Light to deep cherry red, mahogany, brown, or a combination of any of these colors with white, or white alone; distinctive red and white markings preferred.
3. *Size:* Mature cow in milk should weigh at least 1200 lb.

Guernsey

Guernseys have size and strength, and quality and character are desired.

1. *Head:* Clean cut, proportionate to body; broad muzzle with large, open nostrils; strong jaw; large, bright eyes; broad and slightly dished forehead; bridge of nose straight; ears of medium size and alertly carried.
2. *Color:* Shade of fawn with white markings throughout clearly defined; when other points are equal, clear (buff) muzzle is favored over smoky or black muzzle.
3. *Size:* Mature cow in milk should weigh at least 1150 lb.

Jersey

Jerseys have sharpness and strength that indicate productive efficiency.

1. *Head:* Proportionate to stature showing refinement and well-chiseled bone structure; face slightly dished; well-set, dark eyes.

2. *Color:* Some shade of fawn with or without white markings. Black muzzle encircled by a light-colored ring; tongue and switch may be either white or black.
3. *Size:* Mature cow in milk should weigh about 1000 lb.

Brown Swiss

Brown Swiss are strong and vigorous, but not coarse. They have size and ruggedness with quality desired. Extreme refinement is undesirable.

1. *Head:* Clean cut, proportionate to body; broad muzzle with large, open nostrils; strong jaw; large, bright eyes; broad and slightly dished forehead; bridge of nose straight; ears of medium size and alertly carried.
2. *Color:* Solid brown varying from very light to dark. Black muzzle encircled by mealy colored ring; tongue, switch, and hooves are black.
3. *Size:* Mature cow in milk should weigh 1500 lb.

Holstein

Holsteins have rugged, feminine qualities in an alert cow possessing Holstein size and vigor.

1. *Head:* Clean cut, proportionate to body; broad muzzle with large, open nostrils; strong jaw; large, bright eyes; broad and moderately dished forehead; bridge of nose straight; ears of medium size and alertly carried.
2. *Color:* Black and white or red and white markings clearly defined.
3. *Size:* Mature cow in milk should weigh a minimum of 1500 lb.

Milking Shorthorn

Milking Shorthorns are strong and vigorous but not coarse.

1. *Head:* Clean cut, proportionate to body; broad muzzle with large, open nostrils; strong jaw; large, bright eyes; broad and moderately dished forehead; bridge of nose straight; ears of medium size and alertly carried.
2. *Color:* Red or white or any combination.
3. *Size:* Mature cow should weigh 1400 lb.

DAIRY BULL UNIFIED SCORE CARD[2]

General Appearance (55 Points)

Attractive individuality with masculinity, vigor, stretch, scale, and harmonious blending of all parts with impressive style and carriage.

(8) *Breed Characteristics*—[see pages 17, 18]

(8) *Stature*—Height including moderate length in the leg bones with a long bone pattern throughout the body structure.

(8) *Front End*—Adequate constitution with strength and dairy refinement.

Shoulder Blades and Elbow—Set firmly and smoothly against the chest wall and withers to form a smooth union with the neck and body.

Chest—Deep and full with ample width between front legs.

(8) *Back*—Straight and strong.

Loin—Broad, strong, and nearly level.

Rump—Long, wide, and nearly level with pin bones slightly lower than hip bones.

Thurls—High and wide apart.

Tail Head—Set nearly level with topline and with tail head and tail free from coarseness.

(23) *Legs and Feet*—Bone flat and strong. Front legs straight, wide apart, and squarely placed.

Hind Legs—Nearly perpendicular from hock to pastern from a side view and straight from the rear view.

Hocks—Cleanly molded and free from coarseness and puffiness.

Pasterns—Short and strong with some flexibility.

Feet—Short and well rounded with deep heel and level sole.

Dairy Character (25 Points)

Angularity and general openness without weakness, freedom from coarseness.

Neck—Long, lean, and blending smoothly into shoulders; clean-cut throat, dewlap, and brisket.

[2]This score card is reprinted with the permission of the Purebred Dairy Cattle Association.

Withers—Sharp with chine prominent.

Ribs—Wide apart; rib bones wide, flat, and long.

Thighs—Incurving to flat and wide apart from the rear view.

Skin—Thin, loose, and pliable.

Body Capacity (20 Points)

Relatively large in proportion to size and age, providing ample capacity, strength, and vigor.

Chest—Large, deep, and wide floor with well-sprung fore ribs blending into the shoulders; crops full.

Body—Strongly supported, long, deep, and wide; depth and spring of rib tending to increase toward the rear.

Flanks—Deep and refined.

PRACTICAL APPLICATION

Show-ring judging or actual selections on the farm are not accomplished by the aid of a score card. However, the score card serves as a guide, and a complete knowledge of it will be useful both to the beginner and to the experienced judge.

In actual judging, the values assigned to points on the score card are not always used to arrive at a final decision. The degree of deviation from the ideal must be given careful consideration. For example, feet and legs are assigned a total of 15 points. A particular animal with very bad feet and legs which impair the function and actually shorten the life of the animal will be discriminated against much more than indicated by the 15 points on the score card. The characteristic is so deficient that it does not permit complete use of the other parts of conformation. In this case, the final judgment is based on the various conformation characteristics, as they affect the productive capacity of the cow.

4
COMPARATIVE JUDGING TECHNIQUES AND TYPE STANDARDS

Many of the fundamentals of successful comparative judging are given in the Preface of Chapter 1. In brief, comparative judging refers to ranking a number of individuals in order of preference according to type, or, in other words, show-ring judging. As has been pointed out, this judging must always emphasize the practical considerations and accomplishments in breeding a more useful cow. Beauty of conformation is not the only factor. Good livestock breeders love their cattle and work associated with them. This devotion is often reflected in a better performance. At any rate, the emphasis must continually be placed on the functional traits of greatest practical importance. These are:

1. Udder conformation and attachments, especially the suspensory ligament.
2. Dairy character with strength.
3. Feet and legs.
4. Size, stature, and substance with quality.

The first requirement for uniform results from show-ring judging and herd classification is to have a crystal-clear type standard on which to base decisions and selections. The same principle applies to a breeder who wishes to have a game plan for a balanced breeding program of high production and outstanding functional type. The aim need not necessarily be slanted toward show cattle, but the objective should be a herd of sound cattle with good wearing and working traits. The breeder needs an efficient milk-producing machine that requires a minimum of expensive veterinary care. A type standard for a cow of each breed will be described in this chapter. Later, in the chapter on heifer judging, consideration will be given to the kind of heifer suggested as a recommended standard for a mature cow.

Once a type standard has been established, the judging procedure and mental process involved in reaching a sound decision proficiently include the following steps:

1. Observe (analyze).
2. Evaluate (degree of difference).
3. Decide (based on practical importance).
4. Describe (clear, concise reason to justify judgment).
5. Defend (explain intelligent choices to represent values and priorities).

Confidence and judgment in making sound decisions are developed by observing enough good cattle to acquire impressions for a standard. This is supplemented by looking at many cattle in good practical herds to form an opinion on traits that prevail in the breed population. Soon you will develop the technique of looking at a cow and seeing many traits at one glance and forming an immediate impression of the strengths and weaknesses of an individual. You will soon develop to the point where you can look at a cow and determine where a particular individual needs fixing or for what traits she is exceptionally strong. This practice combined with the initiative, spirit, self-reliance, mental discipline, special powers to concentrate, and other constructive developments are the foundation of good judging procedure. Especially important is the ability to see many things at a glance and to make a mental picture of the observations for recall later.

IDEAL BREED TYPE STANDARDS

A successful judge is thoroughly familiar with every detail of the correct or ideal type standard. He or she evaluates and selects according to the breed standard. Figures 14 through 25 show outstanding breed type standards and (or) outstanding cows of each breed.

Type refers to an ideal or standard of perfection that combines all the body characteristics that contribute to the usefulness of dairy animals. All of the cows pictured for the various breeds have these useful characteristics, as indicated in their similarity of good udders, strong frames with plenty of refinement, dairy quality combined with smoothness, straight toplines, and good legs which fit the rest of their bodies. In addition to these common denominators of type possessed by the cows of all six breeds, they have definite breed individualities.

Breed type includes the desirable characteristics of conformation men-

Figure 14 This twice All-American Aged Ayrshire cow's conformation was so generally accepted for excellence that the Ayrshire Board of Directors authorized the use of her photo as representing closely the ideal for the breed. She was officially scored Ex 96.9. (Courtesy Pinehurst Farms and the Ayrshire Breeders Association. Sheboygan Falls, WI, and Brandon, VT)

Figure 15 This 3E 94 cow at 14 years, 9 months received 7 All-American nominations and twice All-American, at 11 and 13 years, and three successive years Reserve All-American. Her tremendous dairy character and excellent udder are obvious in this photo taken at 11 years, 3 months of age. She died at 15 years, 1 month. During eleven lactations of 305 days she averaged 15,514 lb milk, 4%, and 620 lb fat. Her lifetime production was 171,800 lb milk and 6864 lb fat. She has six sons in AI. (Courtesy Mary Shank Creek, Palmyra Farm, Hagerstown, MD)

tioned above, plus specific characteristics that distinguish one breed from another, including color, size, shape, style, and many other traits. These characteristics are so definite that the experienced judge only needs to see the outline of an individual to identify the breed. The term *breed character* is used to describe this distinctiveness in conformation. Some breeds are more distinctive than others, and hence they show more specific breed character.

Ayrshire Character

The emphasis in breed character in the Ayrshire is on a specific type which the inexperienced judge often does not recognize. This very important breed character consists of alertness, style and symmetry, clean-cutness, and sharp features throughout with shapely, evenly balanced, strongly attached udders.

Brown Swiss Breed Type

Over the years the Brown Swiss breed has made great strides in changing from a very rugged, somewhat coarse individual with a mediocre udder to a very attractive cow displaying marked dairy character which combines dairy refinement with scale and size into a suitable machine for uniformly high milk production and a high lifetime accomplishment. It has been the aim to combine sufficient strength with dairy quality in order to avoid individuals which, because they lack scale and a strong frame, are too frail to withstand the strain of year-after-year production.

Brown Swiss have more strength, especially strength of bone, in their type than any other dairy breed. As a group they have easily the most desirable set and strength of legs of all the dairy breeds. Brown Swiss breeders, judges, classifiers, and breed association representatives have all tried to maintain this strength of leg in the breed by discriminating against cattle that do not possess the good leg and feet characteristic. A research study for the dairy breeds in Iowa showed that the herd life expectancy is slightly higher for cows of the Brown Swiss breed.[1]

Guernsey Breed Type

The Guernsey breed is generally recognized as displaying the most dairy character (sharpness, cleanness, and openness of ribbing) of the dairy breeds, although in many respects it is close between the Guernsey and Jersey breed. Traditionally, the average classification scores on dairy character for Guernseys are very high. This indicates that the breed is uniform with outstanding dairy character. Seldom does one see the fat Guernsey cow that was so prevalent at one time. It should

[1]*Journal of Dairy Science,* 53 (1970), 764.

Figure 16   The true type model of a Brown Swiss cow. Note the dairy quality, outstanding mammary, and other features combined with a strong frame and sound feet and legs. (Courtesy Brown Swiss Breeders Association, Beloit, WI)

Figure 17   An outstanding Brown Swiss cow displaying type characteristics indicative of her high production (+ 6270 milk, + 28 fat from herdmates) with All-American type conformation. This cow won many national shows and was voted All-American for 5 consecutive years starting with the 3-year All-American Award. She was voted unanimous All-American two years in succession. She was also selected as the All-Time All-American Aged Cow and 4-Year-Old-Cow as well as the Reserve All-Time All-American 3-Year-Old Cow of the Brown Swiss breed. (Courtesy Bernard Monson, Gowrie, IA)

Figure 18 True type Guernsey cow suggested for a model of the breed. Note the dairy character combined with upstandingness, strength, and size. (Courtesy American Guernsey Cattle Club, Peterborough, NH)

Figure 19 This Ex 94 Guernsey cow was All-American 3-year-old, Reserve All-American 4-year-old, and twice All-American Aged Cow. She was so outstanding that she was named Supreme Champion of All Breeds at the World Dairy Exposition. (Courtesy of Lester Lurvey and Sons, Baraboo, WI; John B. Merryman, Sparks, MD; and Chester and Barbara Williams, Waukesha, WI)

be mentioned that quality, including sharpness and freedom from excess fat accumulation, is given an important place in the specific Guernsey breed type. General refinement with plenty of strength, together with the proper kind of head and neck, is also emphasized to give the Guernsey individuality in breed type.

A special effort has been made to breed and recognize a larger, more up-standing cow for the Guernsey breed.

Holstein Type Standard

The characteristic scale with the proper degree of dairy refinement is important in the Holstein breed. It is one of two large breeds. Size is emphasized in order to have a suitable machine for uniformly high milk production.

The Holstein breed has had an enviable record for exceptionally high milk records for individual cows. In order to maintain this breed characteristic, it is best to have strength without sacrificing too much dairy character and quality. If an individual lacks scale and a good strong frame, she is too frail to withstand the strain of year-after-year production. If she is too coarse, she lacks dairy quality; the odds are against her reaching and maintaining the high peak of production expected from the larger breeds. Oversized animals require too much feed. In order to produce efficiently, the extremely large cow must meet production requirements beyond the expected average. In addition, the oversized cow needs larger than normal housing space. For these reasons, the breed has a specific size that is considered correct.

Although it is not necessary to emphasize breed character of Holsteins as much as for other breeds, they have a distinct type. In denotes a large well-framed cow, with ample strength and depth of body to enable her to consume large quantities of feed. Breeders prefer to have this strength combined with a definite size, smoothness, balance, and blending of parts, together with a straight top, good udder, and desirable legs.

Jersey Character

Desirable breed character in Jerseys depends on refinement and quality starting with a well-dished head displaying sharp, well-defined features. This general pattern of refinement and superb dairy quality is followed throughout the body conformation, but it is combined with strength. Freedom from coarseness and excess fleshing is very important. Unsightly fat deposits on hips, pins, and other part are not removed as readily from Jerseys as they are from the larger breeds.

Upstandingness and strength are important characteristics of Jersey type. These and the special refinement possessed by the Jersey, combined with a deep body and good udder, are very characteristic of the breed. If Jerseys are too coarse and rugged, they lose their best qualities. The Jersey breeders are concerned with high levels and efficiency of production.

Figure 20 The aged cow in this picture represents the epitome of a breeding program for production, type classification, and show record achievements. This 3E 97 Elevation daughter was voted unanimous All-American Aged Cow after she was Grand Champion Holstein and Supreme Champion for all breeds at the Central National Show. She repeated as All-American Aged Cow the following two years. Previously she was All-American 3-year-old. Later she was voted All-Time All-American Aged Cow and Reserve All-Time All-American 3-year-old. Her production records on 2X in 365 days were: 7 years, 7 months, 48,731 lb milk, 4.2%, 2028 lb fat, and 44,143 lb milk, 3.9%, 1698 lb fat at 5 years, 9 months. She was first on the Index listing with a CTPI + 1088. (Courtesy Woodbine Farms and Romandale Farms Ltd., Arville, PA and Unionville, Ontario)

Figure 21 This Holstein cow represents the ultimate in accomplishment in a breeding program stressing both production and type. This cow was Reserve All-American during her first four lactations and then All-American for two successive years. She received High Honorable Mention for All-Time All-American Aged Cow. She is officially classified 5E 97, has four records of over 30,000 lb of milk, has a lifetime total of over 225,000 lb of milk with 4% fat, and has been identified as "The Ultimate Dairy Machine." (Courtesy Ron Hetts, Fort Atkinson, WI)

Figure 22 The ultimate in Jersey type is displayed by this outstanding Ex 97 Jersey cow. She was sired by a bull having a high predicted difference for milk production. She has produced up to 25,010 lb of milk and 1119 lb of fat and over 200,000 lb of milk and 9,000 lb of fat in her lifetime. She had three consecutive records well over 20,000 lb of milk at 4 years, 1 month, at 5 years, 2 months, and at 6 years, 3 months of age. She was Grand Champion and Best-Uddered Aged Cow at the National Jersey Show and later was designated All-American Aged Cow. (Courtesy Happy Valley Farm, Danville, KY, and Briggs and Beth Cunningham, Newberry, SC)

Figure 23 This Ex 96 Jersey cow was named All-American Aged Cow for 3 consecutive years and was twice Grand Champion of the National Jersey Show. Note her extreme dairy character and depth and openness of rib. (Courtesy Waverly Farm, Clearbrook, VA)

Milking Shorthorn Breed Type

In 1969 the Milking Shorthorn breed changed in name from a dual-purpose breed to a dairy breed. The breed has rapidly increased in milk production and the type standards have changed to cattle having a more clean-cut, angular appearance. The breed emphasizes strong but not coarse individuals and good size.

The fundamentals for evaluation of type are much the same for the various breeds, as mentioned in the discussion of the unified score card. However, considerations of breed type as discussed in this chapter must be included for comparative judging. Allowance for this is made on the score card by the points assigned to breed character in the General Appearance section.

IMPORTANCE OF PROPER EVALUATION

In comparative judging each individual must be properly analyzed and evaluated before comparisons can be made. The correct analysis of an individual is based on the proper evaluation of each part of the cow as compared to an established type standard. The judge cannot evaluate properly unless he or she

Figure 24   This All-American Milking Shorthorn Mature Cow exhibits excellent strength and a well-attached, milky udder. Her first four records averaged 14,275 lb of milk and 559 lb of fat. (Courtesy Elmer Von Tungeln, Verden, OK)

Figure 25 This All-American 3-year-old Milking Shorthorn is an example of a fine animal of the breed. She has excellent breed character about her head and throughout. She is also outstanding in dairyness and possesses a very shapely udder. Her 2-year-old record was over 15,000 lb of milk and 600 lb of fat. (Courtesy Kingsdale Farm, Oneida, IL)

has an accurate knowledge of not only what constitutes the correct type for each part of body conformation but also all deviations from this established standard. To help the beginner acquire the skill of systematic analysis and evaluation, the next nine chapters are devoted to a discussion of desirable and undesirable features of the more important parts of conformation. Once these are clearly understood, the analysis of an individual becomes easy, and a rank in a class can be established without any difficulty. When this technique is used either for teaching or for judging, a correct procedure is assured. If students and other judges train themselves to analyze the individuals in a class properly, before ranking them, they will make fewer mistakes and be more successful in their judging.

The best way to learn judging is to master the correct analysis of each individual at the outset. The placing is often evident after the correct analysis and the proper evaluation of the individual animals have been made. This technique also makes it easy for the judge to give accurate, detailed reasons to satisfy participants in judging competition, exhibitors in the show ring, and spectators.

5

EVALUATION OF DEFECTS

One of the first procedures in judging a cow should be a check for the presence of any defects, since these may have a marked influence on the placing. If defects are severe or constitute a disqualification, they should be the deciding factor in determining the placing. Thus such characters are properly given first consideration among the parts of conformation.

A serious defect may be defined as a gross fault that impairs productive performance and/or is so important from the hereditary standpoint that the judge must be very critical in his or her evaluation. Some are considered serious enough to constitute a disqualification, which means that the animal is not eligible to win a prize or to be shown in the group classes. If it seems advisable to rank a disqualified animal in a class, all the other individuals must be placed before the disqualified one is ranked. If several disqualified animals are in the same class, they are ranked on their individual types after all the sound animals have been assigned a placing.

Deficiencies and defects may range in discrimination from slight to serious. It is the responsibility of the judge to determine the degree of discrimination and to assign the proper evaluation to each point of conformation under consideration. In evaluating a specific condition, it is usual to consider its functional importance and whether or not it is inherited.

A list of defects, published by the Purebred Dairy Cattle Association, has gained wide acceptance and is generally used as a standard in judging.

EVALUATION OF DEFECTS IN DAIRY COWS

Permission has been obtained from the Purebred Dairy Cattle Association to reproduce the following evaluation for a dairy cow:

HORNS: No discrimination for horns.

EYES:

1. Blindness in one eye—slight discrimination.
2. Cross or bulging eyes—slight discrimination.
3. Evidence of blindness—slight to serious discrimination.
4. Total blindness—disqualification.

WRY FACE: (Face twisted sideways—see Fig. 26.) Slight to serious discrimination.

PARROT JAW: (Overshot jaw—see Fig. 27.) Slight to serious discrimination.

SHOULDERS WINGED: Slight to serious discrimination.

TAIL SETTING: Wry tail (see Fig. 28) or other abnormal tail settings—slight to serious discrimination.

CAPPED HIP: No discrimination unless affects mobility.

LEGS AND FEET:

1. Lameness—apparently permanent and interfering with normal function— disqualification. Lameness—apparently temporary and not affecting normal function—slight discrimination.
2. Evidence of crampy hind legs—serious discrimination.
3. Evidence of fluid in hocks—slight discrimination.
4. Weak pastern—slight to serious discrimination.
5. Toe out—slight discrimination.

Figure 26 A wry face, twisted sideways. (Courtesy University of Minnesota, St. Paul, MN)

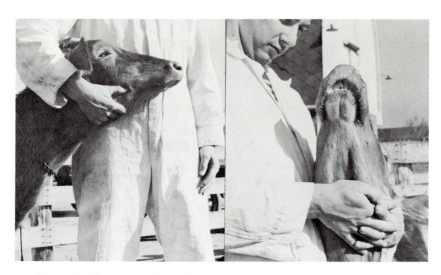

Figure 27 Parrot jaw, also referred to as overshot jaw. Side view and ventral view of jaw on yearling heifer. (Courtesy L.O. Gilmore and H.E. Kaeser, Columbus, OH)

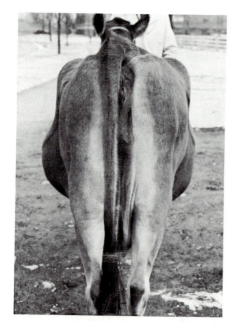

Figure 28 A wry tail may mean the tail head is set either to the left or to the right. (Courtesy L. O. Gilmore, Columbus, OH)

UDDER:
1. Lack of defined halving—slight to serious discrimination.
2. Udder definitely broken away in attachment—serious discrimination.
3. A weak udder attachment—slight to serious discrimination.
4. Blind quarter—disqualification.
5. One or more light quarters, hard spots in udder, obstruction in teat (spider)—slight to serious discrimination.
6. Side leak—slight discrimination.
7. Abnormal milk (bloody, clotted, watery)—possible discrimination.

LACK OF SIZE: Slight to serious discrimination.

EVIDENCE OF SHARP PRACTICE (See PDCA Show Ring Code of Ethics, page 308):
1. Animals showing signs of having been tampered with to conceal faults in conformation and to misrepresent the animal's soundness—disqualification.
2. Uncalved heifers showing evidence of having been milked—slight to serious discrimination.

TEMPORARY OR MINOR INJURIES—Blemishes or injuries of a temporary character not affecting animal's usefulness—slight discrimination.

OVERCONDITIONED—Slight to serious discrimination.

FREEMARTIN HEIFERS—Disqualification.

EVALUATION OF DEFECTS IN DAIRY BULLS

The Purebred Dairy Cattle Association has published an evaluation of defects in dairy bulls corresponding to the evaluation for dairy cows. In general, the majority of the points receive the same discrimination except that a bull with one testicle or with abnormal testicles is disqualified.

SUMMARY OF DISCRIMINATIONS

Disqualifications:

1. Total blindness.
2. Permanent lameness.
3. One or more blind quarters in cows.
4. Only one testicle or abnormal testicles in bulls.
5. Evidence of sharp practice.
6. Freemartin heifers, unless proved pregnant.

Serious Discriminations:

1. Strong evidence of blindness.
2. Wry face (marked).
3. Parrot jaw (pronounced).
4. Badly winged shoulders.
5. Very abnormal tail setting.
6. Very weak pasterns, extreme toeing-out in rear (Fig. 29), badly bowed pasterns (Fig. 30), or a marked spread of toe (Fig. 31).
7. Crampy hind legs.
8. An extreme lack of size.
9. Very abnormal milk or a partially impaired quarter.
10. Broken udder attachment.
11. One or more very light quarters, hard spots in udder, or obstruction in teat (spider).
12. Lack of defined halving with deep udder and/or strutting teats.
13. Severely overconditioned.
14. Uncalved heifers showing evidence of having been milked.

Slight Discriminations:

1. Blindness in one eye.
2. Cross or bulging eyes.
3. A slight tendency toward a parrot, or overshot, jaw in a female.

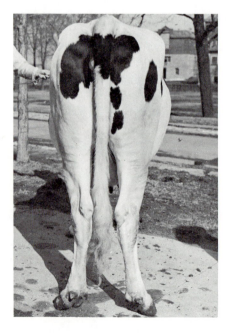

Figure 29 Severe toeing-out of the rear legs places additional strain on the pasterns, which are considered a vulnerable structure. (Courtesy L. O. Gilmore, Columbus, OH)

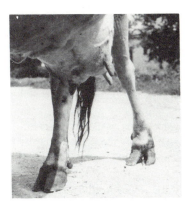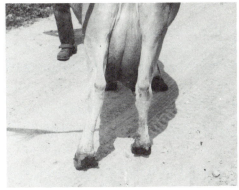

Figure 30 Bowed pasterns in the rear legs are caused by unequal growth of the phalanges, as indicated in the front and rear view pictures. It may appear in different degrees and must be evaluated accordingly. (Courtesy Atkeson, Eldrige, and Ibsren, Manhattan, KS)

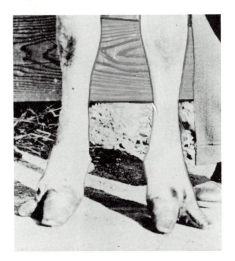

Figure 31 Spread toes on either the front or the rear feet may cause serious foot trouble if the spread is pronounced. (Courtesy S. W. Mead and others, California Agricultural Experiment Station)

4. Cropped ears.

5. Loose shoulder attachment with a slight tendency to wing.

6. Slightly wry tail or other slight deficiency about the tail setting.

7. Temporary lameness.

8. Evidence of fluid in hocks.

9. Slightly undersized.

10. Temporarily abnormal milk.

11. A tendency toward weakness in udder attachment.

12. Slightly unbalanced quarters.

13. Temporary or minor injuries that do not affect the animal's usefulness.

14. Slightly overconditioned.

A disqualification places the individual at the bottom of the class. A serious discrimination brings a heavy penalty and has a marked effect on the placing, in contrast to a slight discrimination which has only a small influence on evaluation for the final placing.

It is conceivable that a slight discrimination will not change the placing of a good cow, especially when the class is not very close. But in a very close class, one in which the individuals are of approximately equal merit, a slight discrimination may influence the rank by several places. A serious discrimination usually places the individual in the lower half of the class, and frequently in about the middle of the lower half or three-quarters of the way down in the class. Many factors or differences in conformation must be taken into consideration in arriving at the final placing, but the above summary can serve as an approximate guide.

UNDESIRABLE RECESSIVES

Skeletal defects are functional deformities. Muscular defects are next in importance. Inherent physical abnormalities are of great economic importance. Breed and artificial insemination associations recognize this by identifying carriers of undesirable recessive traits and publishing lists of bulls carrying these traits. Once identified as a carrier of an undesirable recessive factor, the animal may be purged; in some cases, however, breeders may decide to retain the individual and "breed around" the undesirable rather than completely eliminate the animal from the breed.

The Holstein Association has a list of abnormalities that have been defined as undesirable. Included in the list and transmitted through recessive genes are: bulldog, mulefoot, prolonged gestation, dwarfism, pink tooth, hairless and imperfect skin. The association's bylaws call for periodic publishing of a list of all carriers of undesirable recessives and identification with a code whenever the animal's name appears in a pedigree.

The American Jersey Cattle Club (Reference: *Jersey Journal* Vol. 32, No. 8, page 67) recognizes two genetic abnormalities with an official listing of sires that are carriers. These are quoted:

1. Limber Legs—calf has little or no control over movements of legs and is unable to stand. Calf's legs lack normal muscling, appear loose at the joints and can be flexed, extended and rotated without difficulty or discomfort to the calf. Usually the legs can be crossed above the dorsal side of the neck without discomfort to the calf.

2. Rectovaginal Constriction (RVC)—Constriction of the rectum and vagina such that the arm usually may not be inserted normally in the rec-

tum to permit artificial breeding and episiotomy or caesarean section is usually required for calving; may be accompanied by hardening of the udder.

Usually positive identification of parentage is made by blood testing. Frequently a well-bred and potentially valuable bull is tested by mating with known recessive female carriers so he can be listed as recessive free even though he is from a mating of individuals with the recessive trait in the pedigree. A common listing is one such as Recessive Tested Mulefoot Free.

6
HEAD AND NECK CHARACTERISTICS

The head and neck often reveal quite a bit about a dairy animal. Head and neck characteristics may vary considerably among individual cows, but points of strength or weakness in this region often reflect similar characteristics throughout the entire body of the animal. The two cows pictured in Figures 32 and 33 demonstrate this feature. The cow with the stronger head and neck is also larger and displays greater depth of shoulder, a stronger body, and more strength of bone throughout her frame.

Disposition, strength, constitution, and quality are usually expressed in the head. A pleasant disposition and general strength normally guarantee good feeding qualities, which are becoming more important with the present-day methods of handling dairy cattle.

The proper kind of head and neck adds to the conformation of an individual, and many of the specific breed characteristics are carried here. This is recognized on the dairy cow score card, with 5 points assigned to breed characteristics, including the head. The dairy bull score card emphasizes the head and breed characteristics to an even greater extent by allotting 8 points to this from the 55 points designated for general appearance.

A clean-cut, alertly carried head delineates not only breed character but also dairy quality. A long, lean, well-defined, and smoothly attached neck displays true dairy character. A short, blocky head and thick neck shows a tendency to early maturity and fat predisposition which is undesirable in dairy cattle, and they are usually found in an individual with a short, compact, and heavy body. Sometimes this is associated with calving difficulties. As a rule this is associated with low production. A moderately long head usually indicates a large, open-ribbed body conformation; an extremely long head, however, may reflect a weak constitution, combined with too rangy a body and slow maturity. It should be mentioned that an occasional exception may be found to these general tendencies.

THE IDEAL HEAD AND NECK

The head should be of medium length, clean-cut, and strong; the muzzle should be broad with large, wide-open nostrils and muscular lips. These characteristics reflect a strong constitution, good feeding qualities, and a satisfactory respiratory system. A lean, strong jaw indicates proper refinement combined with the strength necessary to consume and masticate the large quantities of feed required for sustained heavy production. Full, bright, but placid eyes with a gentle expression display pleasant temperament, good health, and vigor. A broad, lean, and moderately dished forehead shows breed and dairy character.

Figure 32 The strength, balance, and smoothness displayed in the head and neck of this cow are carried throughout her body conformation. Her great depth of body and general strength, combined with dairy quality, are admirable. This cow had a high lifetime production. (Courtesy Woodacres Farm, Princeton, NJ)

Figure 33 This is a satisfactory head, but it does not display the exceptional strength of power in the head and throughout the body as found in the cow in Fig. 32. Note the difference in strength of frame indicated, especially in the legs.

To this should be added a straight bridge for the nose, a forehead of proper depth and width, with medium-sized ears which are alertly carried to give the correct balance and attractiveness to the head. Rarely are horns present in females in a modern dairy herd. Horns are conducive to injuries to other cattle and the owner. Thus, economics dictates that the horns be removed either by dehorning at the proper age or genetically, by selecting for the polled trait. For this reason, all cows and heifers pictured in this text are hornless. Some of the bulls have horns because they are used in tying, but individuals with horns are dangerous. When bulls are handled in groups, horns are usually removed.

The neck should be long, lean, clean-cut, and strong with a neat attachment of the head on one end and blending smoothly into the shoulders and brisket on the other. The throat and dewlap should be clean-cut, since this adds to the trimness and neatness of the individual. It is very important for the neck to have the correct dairy character. A short, thick, compact, and/or coarse neck usually shows poor dairy character and lack of productive ability.

A description similar to this can be used for bulls except that they are masculine and stronger in the head, have horns of medium size, and carry a crest on a much stronger neck.

DEVIATIONS FROM THE IDEAL

Characteristics of head and neck which deviate from the ideal, with the degree of discrimination indicated in parentheses, include the following:

1. Narrow, weak muzzle (serious).
2. Long, narrow, weak head (slight to serious).
3. Small and restricted nostrils (slight to serious).
4. Small, dull, deep-set eyes (serious).
5. Roman nose (too prominent bridge of nose) (slight).
6. Weak, shallow jaw (slight to serious).
7. Short, shallow forehead (slight).
8. Narrow poll (very slight).
9. Pointed and/or too prominent poll (very slight).
10. Plain head, lacking in dish, sharpness, and breed character (serious).
11. Ears too small or too large (slight).
12. Inherited notched ears (slight).
13. Thick, short head (slight to serious).
14. Large, coarse head (serious).
15. Listless or undignified carriage of head (slight).
16. Short neck (slight).
17. Thick, heavy neck (slight to serious).
18. Too much dewlap, heavy brisket, lack of clean-cutness and refinement (slight to serious).
19. Too prominent a neck, showing crestiness in a female (slight to serious).
20. Ewe neck, lacking somewhat in strength (slight).
21. Neck lacking smoothness of attachment (slight).

BREED DIFFERENCES

The differences in breed character about the head, among the various breeds, should be recognized and given proper evaluation in judging, as has been mentioned under the discussion of breed character for the score card in Chapter 3.

Jersey breed character consists of a well-balanced head, with a slight dish and dark eyes that are well set. During recent years the emphasis in Jersey type has been on a somewhat larger cow, and the recommended weight has been increased by 100 pounds. Simultaneously, a longer head was recommended to coincide with a larger body and to keep the proper balance or proportion.

The pictures that accompany this chapter demonstrate differences in breed character about the head and neck far better than is possible by description. A careful study of these pictures is recommended.

The Ayrshire head and neck are clean-cut, refined, but strong, and the alert carriage of the head gives a definite style and flash to individuals of this breed. Sharpness, specific Ayrshire character, strength with dairy quality, and balance are given major consideration. The head is moderately long, with less dish than that of the Jerseys and Guernseys. It takes some study for the beginner to appreciate the typical Ayrshire character and to be able to identify it.

In the Guernsey breed the emphasis is on sharpness, with a display of clean-cutness and dairy quality to coincide with the outstanding dairy character carried throughout the body conformation of this breed.

For judging heads and necks in the three smaller breeds described above, it should be mentioned that coarseness, untidiness, shortness, heaviness, compactness, thickness, lack of proportion, and dullness of eye are considerd very objectionable qualities.

The emphasis on an attractive head in Holsteins, Milking Shorthorns, and Brown Swiss is concerned primarily with a combination of strength and refinement. A head with a proper proportion or balance, and indicating a vigorous constitution, is ideal, but more variations are tolerated here than in the smaller breeds. This is another way of saying that less emphasis is placed on a specific breed character for Holsteins, Milking Shorthorns, and Brown Swiss, and more emphasis is given to qualities that indicate economic usefulness and ensure the uniformly high production of these three breeds. Special attractiveness about the head and neck is recognized but is not given as much weight in the final placing in Holsteins, Milking Shorthorns, and Brown Swiss as in the other breeds.

Figure 34   Three Brown Swiss cows with excellent, strong heads. The cow on the right produced 27,542 lb of milk, 4.8% test, with 1326 lb of fat. Her strength about the head indicates this exceptional production. At the completion of her yearly record, she was milking 64 lb a day. (Courtesy Kilravock Farm, Litchfield, CT)

Figure 35 A pair of strong yet refined Jersey heads typical of the breed. Note the broad muzzles, prominent eyes, and slight dish to the forehead. The cow on the left is the reigning National Milk and Fat Champion; the cow on the right is her maternal sister and was former National Fat Champion. (Courtesy of the American Jersey Cattle Club, Columbus, OH)

Excellent types of Brown Swiss and Holstein for head and neck characters are shown in the accompanying pictures. Both breeds have a rather long head. However, it should be pointed out that the Brown Swiss possess more substance about the head and neck than any other breed. Usually the neck is more developed, and sharpness and clean-cutness are not considered as important here as they are for the other breeds. It is generally known that much more throatiness, somewhat more coarseness, and a heavier dewlap and brisket are tolerated in judging Brown Swiss than in judging any other breed.

Figure 36 Breed and outstanding dairy character are demonstrated by the sharp, well-balanced, powerful heads of these three Brown Swiss cows on a breeder's farm in Iowa. (Courtesy of Bernard Monson, Gowrie, IA)

Figure 37 An Ex 97 Jersey cow pictured with three of her embryo-transfer offspring. All four possess excellent Jersey breed character. (Courtesy of the American Jersey Cattle Club, Columbus, OH)

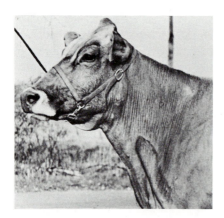

Figure 38 Brown Swiss head showing remarkable clean-cutness and dairy quality. The chiseled features and the strong, deep jaw all indicate structural and constitutional strength. The unusual refinement reflects the great dairy quality and productive capacity of this cow. (Courtesy Walhalla Farms, Rexford, NY)

Figure 39 In contrast to the heads pictured in Figs. 36 through 38, which display outstanding dairy refinement, this head and neck are coarse, heavy, and seriously lack breed and dairy character. As can be expected, they are carried on a heavy, coarse, compact, close-ribbed, and thick-bodied cow.

Figure 40 These young Guernsey heifers show considerable differences in head types. The long, clean, strong but refined head on the heifer in the middle is very outstanding. (Courtesy American Guernsey Cattle Club, Peterborough, NH)

Figure 41 Dairy and Guernsey breed character are displayed by the two curious young Guernsey cows. (Courtesy Cornell University, Ithaca, NY)

Figure 42 A parrot jaw, indicated by the very long, pointed nose and lack of strength in the forepart of the lower jaw. This cow made a number of satisfactory records. A moderate discrimination for this case.

Figure 43 The attractive head on this Guernsey cow shows not only tremendous dairy character but also strength and charm. She has an enviable show record of winning All-American Aged Cow awards 5 years in a row, has classified excellent eight times with 94.5 the highest score, and has consistently produced about 14,000 lb of milk. She has three consecutive records of over 700 lb of fat in 305 days on 2X. (Courtesy Woodacres Farm, Princeton, NJ)

Figure 44 Excellent head of a Guernsey cow showing breed character, strength, and quality. (Courtesy Cornell University, Ithaca, NY)

Figure 45 This Guernsey head has too "pretty" and dainty a face, a narrow muzzle, and weak jaw. When compared with the cows shown in preceding figures, it is apparent that this is far from the ideal for a Guernsey cow. (Courtesy American Guernsey Cattle Club, Peterborough, NH)

Figure 46 The head of this cow is seriously lacking in strength and depth of jaw. She does not have the strength and vigor to keep producing over a long period of time. The entire frame and structure are too fine to meet the standards of desirable Guernsey type. (Courtesy American Guernsey Cattle Club, Peterborough, NH)

Figure 47 This head is too plain, coarse, and lacking in Guernsey breed character.

Figure 48 Two views of an excellent cow having a score of 94 but showing a plain head and a too prominent Roman nose. This demonstrates that the head does not exert much influence on overall placing or classification.

Figure 49 The head on this newborn calf appears to resemble that of the dam with improvement in the muzzle. The dam has a satisfactory head with considerable width and dish plus a deep well-balanced forehead, but the muzzle is not particularly strong, and the nostril is too restricted. The calf appears superior in these respects and has a very attractive head. (Courtesy Cornell University, Ithaca, NY)

Figure 50 The heads of these two calves show considerable differences. Note the shallow, receding forehead on the first calf from the left. The second calf's head is much deeper from the eyes to the poll and has considerably more overall length. (Courtesy Cornell University, Ithaca, NY)

Figure 51 This Holstein shows splendid strength in her head. Note the width between the eyes, the strong nose, and the great depth of jaw. There is no sign of coarseness to achieve this strength, and the nostril is large and open. The neck is clean and strong. This good head displays only moderate character, and the one in Fig. 53 far surpasses it in this respect. (Courtesy Bob and Wendy Miller, Huntley, IL)

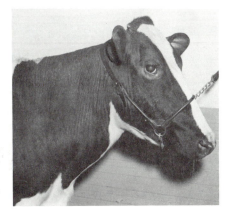

Figure 52 A very weak head that is attached to a weak, underdeveloped neck. The entire structure is too frail for the Holstein breed. Note the shallow forehead; the shallow, weak jaw; the dull eye; the narrow, weak, and very restricted nose; and the small muzzle. A direct opposite of the head is shown in Fig. 51.

Fig 53 A strong, powerful Holstein head having a great deal of refinement and a tremendous amount of breed character. The dish in the forehead, the balance and proportion together with the full and level poll, the finely featured nose, the wide muzzle, and the deep angle of the jaw all contribute toward a near-perfect head. The neck shape similarly denotes strength and refinement. The fine features with strength and character are exceptional. (Courtesy of Holstein–Friesian Association of America, Brattleboro, VT)

Figure 54 This sharp and powerful head exemplifies Ayrshire character in every respect. A study of the details is much more impressive than can be accomplished by description. (Courtesy Gordon Huntington, Amherst, NH)

Figure 55 An outstanding Ayrshire head but not nearly as good as the one in Fig. 54 which had an advantage in strength and length of head and smoother blending at the withers.

Figure 56 Jersey character about the head and neck and throughout the entire conformation is always something special in the good ones. Note the dish, the balance, and character combined with strength for distinctive Jersey breed type. Compare this to the type in Fig. 57.

Figure 57 The shorter, heavier neck, the short distance from eye to poll, the coarseness of jaw, and the plainness about the forehead are a direct opposite to the outstanding Jersey character shown in Fig. 56.

Figure 58 Exceptional Guernsey breed type is possessed by this frequent Grand Champion winner at state, national, and international shows. Note the strength, smoothness, character, and refinement of this head. (Courtesy McDonald Farm, Cortland, NY)

Figure 59 This Guernsey bull, in contrast to the bull shown in Fig. 58 has a plain, coarse head, shows too much loose skin in the region of the throat and dewlap, and has too much crest on the neck, which is heavy and out of proportion to the rest of the body.

Figure 60 The clean-cutness, smoothness, and the Holstein breed character portrayed in this head denote strength and dairy refinement. The balanced conformation and blending of all parts are unusual and add to the attractiveness of the front end of this bull.

Figure 61 Side view of a fine Jersey bull head. Note the balance, style, strength, and dish of forehead, with prominent eye, all of which spell Jersey character. The long, smooth neck possesses ample masculinity without a suggestion of coarseness. (Courtesy American Jersey Cattle Club, Columbus, OH)

7
SHOULDER CONFORMATION

The proper kind of shoulder adds a great deal to the smoothness and strength of a dairy cow or bull. Weakness in the shoulder region can affect the gait of an individual, especially the ease of movement of the front legs. This may sometimes become serious enough to interfere with practical functions, particularly the ability to walk long distances and to spend a sufficient amount of time feeding.

Certain shoulder weaknesses cause cows to tire excessively under difficult conditions, and they become less efficient in their producing ability, especially during advanced age. In contrast, a strong, smooth-shouldered cow can move about freely and can produce efficiently at high levels with a minimum of care. As the cow ages, these differences in wearing qualities become more pronounced, and smooth-shouldered cows usually show the beneficial effects of their good conformation.

For a thorough examination of the shoulder region, it should be observed from the direct front view, a side view, and the back view, while the animal is in motion and while at rest. In addition, observation of the leader at a show often discloses weaknesses of shoulder in the animal he or she is leading. If the leader frequently kicks the shins of the animal, it is usually a sign of "laziness" or "settling" at the point of the shoulder while standing.

THE IDEAL SHOULDER

The ideal shoulder in a dairy cow is reasonably sharp, clean cut, and smooth at the withers or top of the shoulder. This can be observed from the back view and partially from the side. The shoulder blades should be smoothly set against the chest wall and withers so they form a neat junction with the body and neck. The shoulder blade should be long and show considerable depth from the top of the withers to the point of the shoulder. The blade should gradually widen toward the top and the entire structure should be smoothly laid in. The point should be refined and smooth so it blends properly and is not very prominent. Equally important is the attachment of the shoulder, which should be firm, strong, and snug, to allow the shoulder blade to move smoothly while the cow is in motion and also to prevent the point of the shoulder from showing a loose attachment or in extreme cases the wing-shouldered condition while the cow is standing.

The shoulders pictured in Figures 62, 64 through 67, 70, 71, 77, 80, 84, and 87 are close to the ideal. Since no distinct breed differences are indicated in the shoulder, it was not considered necessary to include pictures for all breeds. The same description can be used for shoulders of bulls as for cows, except that the former are more massive and are expected to show the masculinity and substance natural for the male sex. Figures 90 and 92 show nearly ideal shoulders for bulls of the light and heavy breeds, respectively.

DEVIATIONS FROM THE IDEAL

Less perfect conformations of shoulder include the following: heavy or open withers lacking in smoothness; shallow shoulder from the withers to the point; shoulders that are rough or poorly attached; heavy beefy shoulders; coarse, prominent points of shoulders; loosely attached shoulders, to the extent that the animal walks poorly and stands wing-shouldered (or in a milder form shows the point of shoulder too prominently while walking and settles in the region of the point when the animal is not in motion). These characteristics are pictured at various stages in descending order.

It is also possible to have a shoulder with too much refinement, which makes it too frail and lacking in proper strength. This may reflect a lack of strength in the entire frame of the animal and is serious in the larger breeds, which should possess a strong, rugged frame.

In evaluating deviations from the idea, due regard should be given to age and stage of lactation. A cow of advanced age need not be so firm in shoulder attachment or so smooth in the entire region as a young cow.

A dry cow, or one advanced in stage of lactation, need not be as sharp at the withers and can carry more covering over the shoulder than the cow at the peak of her production. A good livestock judge recognizes that changes do take place during the lactation period and will not penalize a cow for this normal condition during the dry period or toward the latter part of her lactation.

The worst fault in shoulder conformation is the winged shoulder, especially in an advanced stage at which the shoulder attachment is almost completely broken and no longer provides the necessary support. The degree to which the individual displays the winged condition determines the emphasis placed on it, but it ranges from a slight to a serious discrimination. The characteristic is considered much more serious in young cows than in those of advanced age that have a right to show some wear. Age should always be considered before deciding on the degree of discrimination for penalizing a cow with winged shoulders. The condition is inherited and often runs in families. It is much more common in some breeds than in others but is present to a limited extent in all breeds and should receive attention in a breeding program.

In giving reasons, terminology of favorable and unfavorable shoulder type characteristics will be useful for properly describing the condition. Some that may apply to certain situations follow:

Favorable:
1. Sharp (clean-cut) and smooth over the withers.
2. Smoothly laid-in at the withers.
3. Well-defined at the withers.
4. Neat and smooth of shoulder.
5. Neatly laid-in at the shoulders.
6. Neatly laid-in shoulders that blend smoothly with the neck and body.
7. Shoulder blades set smoothly against the chest wall and withers.
8. Deep, smooth, and neatly laid-in shoulders.
9. Smoothly laid-in at the shoulder.
10. Blends in smoothly at the shoulder.
11. Shoulders that fit snugly into the body.
12. Smooth at the point of shoulder.
13. Refined and not prominent at the point of shoulder.
14. Strong attachment of shoulder, especially at the point.
15. Nearly ideal shoulders.

Unfavorable:
1. Heavy and coarse at the withers.
2. Opens up at the withers when she walks and stands.
3. Too broad at the withers.

4. Coarse and open at the withers.
5. Heavy at the withers.
6. Thick and rough in the shoulders.
7. Shoulder blades do not blend smoothly into the neck and body.
8. Too open a shoulder.
9. Coarse, heavy shoulder.
10. Too heavy, open, and rough in the shoulder.
11. Heavy, coarse shoulders, which indicate a lack of dairy quality.
12. Shoulder lacks strength.
13. Narrow and frail through the shoulders.
14. Too narrow in the shoulder.
15. A bit shallow in the shoulder.
16. Lacks depth of shoulder.
17. Open and prominent at the shoulder.
18. Very prominent at the point.
19. Out at the point of the shoulder.
20. Extremely loose (weak) in attachment of shoulder.
21. (Slightly), (moderately), (severely) winged shoulder.
22. Broken in shoulder attachment.

Figure 62 A smooth and deep-shouldered Brown Swiss cow. The point of shoulder is excellent. It displays sufficient strength but with such an abundance of refinement that it is hardly visible. (Courtesy Walhalla Farms, Rexford, NY)

Figure 63 Cow with a loosely attached coarse shoulder that is too prominent at the point.

Figure 64 The superb shoulder on this cow reflects her blended conformation throughout. It represents the top in quality and strength. The top of the withers is sharp and smooth. This cow has unusual depth from withers to point of shoulder, which shows remarkable smoothness and blending to provide the extraordinary in symmetry and balance. These shoulders are on an Ex 97 score cow having four records over 30,000 lb of milk and the highest over 1300 lb of fat. She was classified 5E at 15 yr and at that time had 224, 352 lb M, 4.1%, and 9204 lb F. (Courtesy Ron Hetts, Fort Atkinson, WI)

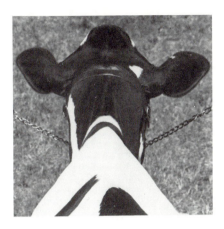

Figure 65 Exceptional back view of cow shown in Fig. 64. She is outstanding in sharpness and smoothness.

Figure 66 Front view of same cow shown in Figs. 64 and 65. It is a rare combination of strength and quality. Note the width of chest but not a sign of coarseness.

Figure 67 The right kind of shoulder on a Junior All-American Senior Calf. Her shoulder shows the same pattern described for the cow shown in Fig. 62. It has the same sharpness on top, great depth from withers to the very smooth point of shoulder, and complete blending of all parts. (Courtesy Brian Derr, Taneytown, MD)

Figure 68 A Holstein heifer having a good, deep shoulder, but she does not have the exceptional smoothness at the point of the shoulder. (Courtesy Cornell University, Ithaca, NY)

Figure 69 A Holstein heifer having rough shoulders indicated by a coarse, prominent point of shoulder that does not blend properly. She is inferior to the calf and heifer shown in Figs. 68 and 67, which are ranked in descending order with this heifer.

Figure 70   A Jersey cow having smooth shoulders that resemble the pattern set up for the ideal in this chapter. She was an All-American Aged Cow. (Courtesy Happy Valley Farm, Danville, KY)

Figure 71   A very smooth shoulder on this All-American and Reserve All-American Holstein Cow. It is a very deep shoulder to correspond with the tremendous scale of this cow. She is decidedly superior in her shoulders to the individuals in Figs. 72 and 73. (Courtesy Apache Ranch, Lapeer, MI)

Figure 72 The good shoulder on this cow is not as smooth and does not blend as well at the point and the upper region of the shoulder, including the withers, as the one in Fig. 71. (Courtesy Cornell University, Ithaca, NY)

Figure 73 A Holstein cow having a rough shoulder indicated by a prominent point of shoulder. The entire shoulder is loosely attached. This is indicated by the position of the point of elbow, which reflects the position of the humerus (Figs. 6 and 7).

Figure 74 This Reserve All-American 4-year-old Ayrshire Cow is very outstanding in every respect except for her shoulders, which are indicated by a loose, prominent point of shoulder. The degree of winged shoulder in this cow suggests a moderate to serious discrimination. She has many strengths to balance against this weakness, but it is also the reason why she did not get the top spot.

Figure 75 A seasoned judge will observe the loosely attached shoulders in this young cow and will predict the slightly winged condition shown in Fig. 76 when the cow is more advanced in age.

Figure 76 This cow shows tiredness and wear in her slightly winged shoulders. It is only a moderate discrimination, and cows do not leave the herd for this condition.

Figure 77 This outstanding Holstein cow shows remarkable sharpness, smoothness, refinement, and strength at the withers and at the point of the shoulder. (Courtesy Collins-Crest Farm, Perry, NY)

Figure 78 An extremely loose shoulder with a coarse, prominent point.

Figure 79 This cow has coarse, heavy, and shallow shoulders.

Figure 80 The shoulders on this 5X All-American Aged Cow show exceptional depth and smoothness.

Figure 81   Strong shoulders having more prominence at the point. These are satisfactory for practical use.

Figure 82   This Guernsey cow shows fine dairy character in the shoulder region, but her shoulder attachment is too loose. This permits the point to settle when she stands and makes it too prominent and loose when she moves.

Figure 83 Another cow having outstanding dairy quality, but her shoulders are so refined and loosely attached that she stands wing-shouldered; the entire shoulder structure moves too much when she walks.

Figure 84 The extreme smoothness and fine type of shoulder displayed by this Holstein cow marked her entire conformation and brought her many grand championships and great success in the show ring. She was designated All-time All-American Aged Cow and later Reserve All-Time, (Courtesy Paclamar Farms, Louisville, CO)

Figure 85 The smooth, strong but refined, neatly laid-in shoulders on this outstanding Holstein cow with high production are reflected with this type of conformation throughout. She displays dairy quality and strength. (Courtesy Charles J. Auger, Winter Place, Winslow, ME, and David E. Paul, Starbright Holsteins, Bath, PA)

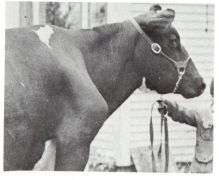

Figure 86 This outstanding cow, the possessor of many high-production records, shows wear in her loosely attached winged shoulders. At an advanced age this fault is not as serious as in a young cow. She was 12 years of age when the photograph was taken.

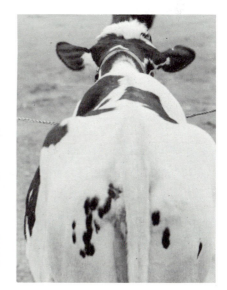

Figure 90 This Guernsey bull has smooth, blending shoulders that show proper refinement, strength, and even contour at the point of shoulder. (Courtesy McDonald Farms, Cortland, NY)

Figure 91 In contrast to the bull shown in Fig. 90, this one has coarse, heavy shoulders that are much too prominent and heavy at the point of shoulder. The coarseness extends forward, especially to the crest of the neck.

Figure 92 This Holstein displays smooth, deep, strong, but refined shoulders.

Figure 93 The heavy shoulders, especially the coarse, rough point of shoulder, of this bull detract from his good conformation in other parts of his body.

Figure 94 The shoulders of this Holstein bull are even poorer than those pictured in Fig. 93. The shoulder conformation is shallow, showing less depth from withers to point of shoulders, and is heavy, prominent, bulging, and coarse at the point. This is partly due to the lack of a strong, smooth shoulder attachment.

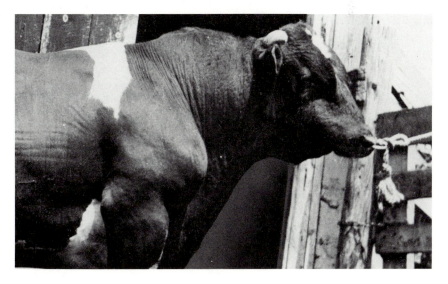

8

JUDGING LEGS AND FEET

Feet and legs and udders are the two most important functional type characteristics. This is true for both the commercial dairy farmer and the purebred breeder. Foot problems are due to both heritability and environment, and the interplay between the two when the susceptibility exists. Attention at the practical level, combined with research findings, has resulted in agreement on what is ideal (correctness of hind legs and feet) and how deviations (deficiencies) should be evaluated.

Some of the heat detection problems are associated with sore feet. Severe foot problems can affect the entire body, especially deep-seated reproductive problems. Losses due to lameness may be in the form of veterinary costs, reduced milk production associated with less feed intake, loss in body weight, lower reproductive performance, and/or replacement costs, including culling due to crampiness or arthritis.

In show-ring judging it is often necessary to penalize an animal that has poor feet and legs to a much greater extent than the 15 points allotted to this characteristic on the score card. This is especially true of young animals that have extremely poor feet and legs, because they seldom develop into useful old animals. For this reason, age should be considered in making a sound evaluation.

The number of good cows that have to be discarded or that gradually decline in general health and production because of bad feet and legs is much larger than is generally realized. Feet and legs can be in such poor condition that they seriously hamper the productive capacity of an otherwise good cow.

Figure 95 Silhouette of legs and feet of a grand champion. (Courtesy Hoard's Dairyman, Fort Atkinson, WI)

Thus a penalty beyond the 15 points allotted on the score card can be well justified. Any breeder who says that he or she has never had to discard a cow because of bad feet and legs is either extremely inexperienced or is pampering such cows beyond practical means.

Conditions under which cattle are kept can markedly affect foot and leg conformation. Wet, swampy land and a muddy yard are conducive to soft, overgrown feet. Proper trimming of the feet (Fig. 97) at regular intervals is im-

Figure 96 The feet and legs in this picture are a poor contrast to the silhouette of those shown in Fig. 95. But they belong to an 18-year-old cow that has had an outstanding lifetime production. Her feet and legs, therefore, are acceptable for a cow at this advanced age. There would be a serious discrimination against such feet and legs on a cow half this age and in the prime of her life.

portant in these situations. The hard outer tissue of the hooves of cows kept continuously on abrasive concrete often wears faster than it grows. This leaves the cow in the situation of walking on a hard surface while bearing most of her weight on the soft part of her hoof. Tender feet and lameness are the result.

Increasing numbers of dairy farmers are changing from grazing to confinement systems of dairying for economic reasons. Almost all confinement systems require cows to spend all or almost all of their time on hard surfaces such as concrete. A major problem with these confinement systems is an increasing number of cows exhibiting foot and leg problems. These problems vary from those caused by too little exercise (hooves not wearing as fast as they grow and resulting in toes growing too long) to those caused by too much exercise (the abrasive surface wearing hooves faster than the hard tissue can be replaced). Other problems include bruised heels caused by improper foot angle or shallow heels, swollen hocks, and weak pasterns. Swollen hocks and weak pasterns often go together. The animals exhibit lameness, which in turn indirectly affects production and reproductive efficiency. The new type card of National Association of Animal Breeders (NAAB) includes separate descriptive areas for feet and rear legs (side view) as well as eight separate foot and leg defects (see Chapter 25). According to reports from various researchers, almost all lameness in dairy cattle is caused by foot and claw abnormalities. Researchers at Cornell (R. W.

Figure 97 Feet can be kept neat and in good shape by proper trimming at regular intervals. Trimming will help maintain correct length of hoof and a strong wall and will prevent the sides from growing under the foot and forming a poor sole or bearing surface. (Courtesy American Guernsey Cattle Club, Peterborough, NH)

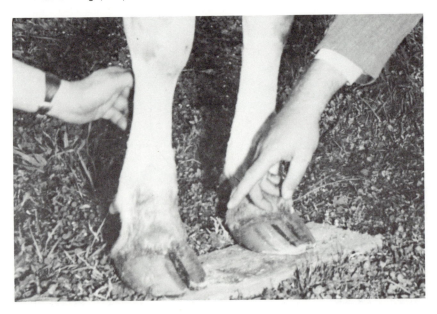

Everett) and at North Carolina (B. T. McDaniel and M. V. Hahn) have found various foot characteristics to be measurable with a high degree of repeatability (hoof length and hoof angle) and others with only a moderate degree of repeatability (heel depth). All characteristics were heritable to a moderate degree. They were also affected by herd environment, such as trimming and walking through a copper sulphate foot bath. Cows in herds that used foot baths had better feet, that is, deeper heels, steeper angles, and shorter claws. Cows continuously kept on concrete or paved lots will show less stress if allowed access to a dirt lot for a few hours a day.

In research work conducted by W. E. Vinson and J. E. Honnette at Virginia Polytechnic Institute, three of the Holstein descriptive type codes for feet and legs were found to be positively associated with herd life and lifetime milk production (code 1—straight, clean, squarely placed; code 2—acceptable; code 4—bone too light). Code 4 was found to have the highest association. Two of the descriptive codes (code 3—close or sickled at hock; code 5—too straight) were negatively associated with herd life and lifetime production. Code 5 (too straight) had by far the highest negative association.

According to current research, cows whose legs are too straight in the hock and that have a low hoof angle and/or shallow heel should receive the most severe discrimination. Converse to earlier thinking, cows having light bone should be penalized less severely, if at all.

CHARACTERISTICS FOR THE IDEAL

A few dairy animals have legs and feet that approach perfection, and these can be used as examples (Figs. 98, 99, 101, 106, 107, 108, 109, 112, 116, 124, 127, 128, 138, 139, 140). Once the ideal is well established in the mind of a judge, he or she can easily detect deviations. These include conditions that impair good movement, appearance, and usefulness.

A proper manner of walking is essential to symmetry and beauty of conformation and displays grace, coordination, and poise. An experienced judge can closely predict the condition of the feet and legs after watching an animal walk. Both the rear and the front legs are important and are described separately. The rear legs, however, are much more important than the front legs and receive the major emphasis.

The leg bone should be strong and smooth, with plenty of substance as well as refinement. This denotes good dairy quality, with bone, tendon, ligament, and muscle construction that combines a substantial foundation with the ability to move easily and smoothly. There should be no indication of roughness or coarseness. To provide good wearing qualities, the leg bone should be hard, flat, flinty, strong, and of a size that is typical and adequate for the breed or weight of the individual (Figs. 107, 108). The hock should be wide and clean when viewed from the side.

The rear legs should have a moderate set, indicated by a slight curve as

Figure 98   An almost perfect leg showing proper curve to the hocks, a strong pastern, and well-formed feet and proper hoof angle. This 3E 97 cow withstood the test of wear on feet and legs to an advanced age. (Courtesy Bob and Kay Miller and family, Mil-R-Mor, Dundee, IL)

illustrated in Figure 98. The correct set, combined with a strong bone, provides just enough spring to the leg to provide smooth, coordinated movement and to carry the weight properly. Thus the hocks and other joints remain clean and free from any unnatural fullness.

The feet should be moderately large with well-rounded toes carried close together and with considerable depth and width at the heel and throughout the foot. When this is combined with a level sole, the foot provides a sound foundation that wears well and requires a minimum of trimming.

Pasterns are an important part of the structure. They should be of medium length and strong but springy. They absorb considerable shock which jars and irritates the joints of the hock and foot with every step. Pasterns should join the hoof smoothly, being properly proportioned, and have the correct curve to give the leg a proper overall contour or shape.

The entire leg structure should be placed squarely under the body to ensure ease of movement (Fig. 101). From the side view, the legs should be almost perpendicular from hock to pastern but also should blend into the curve of the hock and pastern. When viewed from behind, the legs should be wide apart and almost straight (Fig. 112).

The front legs shold be straight and placed with enough width between them to provide ample chest space (Fig. 124). They should move smoothly and, like the rear legs, have the size and strength of bone to fit the rest of the body.

DEVIATIONS FROM THE IDEAL

Probably no other point of conformation has as many degrees of deviation as do the feet and legs. This is due to the complexity of structure involved and to the fact that several parts may change with age and/or under different environment or management conditions.

Depth of Heel

A foot lacking in vertical depth, especially in depth at the heel, becomes flat and needs frequent trimming. It is also subject to other foot troubles because it does not wear well. The shallowness at the heel permits various infections, particularly those likely to cause hoof rot, to gain admittance into the soft, fleshy tissue near the hoof (Fig. 119), creating a great deal of trouble for the owner or manager. Infections require expensive veterinary care, consume valuable time, and frequently reduce milk production, reproductive efficiency, and/or body weight to a marked extent.

A bad foot can exert a great influence on the shape of the leg, especially as an animal ages. By the same token, an ill-shaped leg can have a definite effect on the shape of the foot, on the growth of hoof tissue, on the sole, and on the way the foot will wear. When both the feet and the legs are bad, the ill effects are cumulative and more pronounced. Feet which are small and do not provide enough bearing surface are objectionable. Serious discrimination is given to a condition in which the toes are constantly spread and are carried too far apart (Fig. 31). This usually leads to disorders of the area between the toes and the heel. The conditions under which cattle are kept and the amount of trimming done influence the shape of the foot and the way the feet and legs will wear.

Soft, Spongy Feet

A soft, spongy foot causes a great deal of trouble. Walls that have "curled" under the foot create excess pressure on the sensitive parts of the foot. Pressure on some of the nerve endings in the delicate parts of the foot causes the animal to change the position of the feet, and this produces a severe strain on some of the tendons, ligaments, and muscles. The unnatural pressure at an awkward angle often stimulates excess secretions in tendon sheaths and bursas, which results in a gradual injury to the region, indicated by tender and often inflamed, puffy joints, legs, or pasterns. Unless the condition is corrected at an early stage, a permanent injury which would call for serious discrimination may result. Such penalty is justified because the condition affects the entire well-being and productive capacity of a cow. The endocrine and digestive systems may be affected as well as her general health. Everyone knows the unpleasant effect of tired, aching feet on energy and disposition. The effect on cows is the same. From a practical standpoint, the emphasis placed on feet and legs and the discrimination by judges against bad conformation are well justified because they are of considerable economic importance.

The Pasterns

The pasterns can be too long, too weak, badly shaped, inflamed, or broken (Figs. 102, 117, 118, 119, 120, 121). Since they are such an important part of conformation, there is a serious discrimination for marked deviations from the ideal. Many close placings have been won or lost because of the condition of the pasterns.

The Hock

The hock joint should be wide and strong, but also clean-cut, refined, and cleanly and sharply molded so that it is free-moving and provides ease and grace of movement. Coarse, thick, meaty, weak, or poorly set hocks are objectionable and receive from slight to serious discrimination

Temporarily blemished hocks, due to an injury, are given only a slight discrimination, but badly damaged hocks that appear permanently injured are severely penalized. It is generally known that temporary lameness, which does not affect normal function, receives only a slight discrimination. Permanent lameness, because it interferes with the normal functioning of the individual, is either a serious discrimination or a disqualification.

Too Much Set of Hock, Too Straight, Too Close

Additional defects of the rear hock include the following: sickled or too much set to the rear leg (Figs. 103, 104, 119, 120), post legged or too straight in the hock (Figs. 121, 131, 132, 133) and too close at the hock (Figs. 113 and 114). Since there are varying degrees of these conditions, the discrimination may vary from very slight to very severe. Given the same degree of severity of the fault, and relative to the function of the animal, it is currently thought the most serious of the three problems is too straight in the hock, followed by too close at the hock and then too much set to the hock. A cow with a sickle-shaped leg (showing too much set) is penalized very little if she has a good heel so she does not walk too much on the fleshy part above the heel.

Cows that are too straight in the hock have a shorter herd life and a sizable lower lifetime production. Cows having this condition lack the ability to absorb the shock of normal walking, and the problem is increased when they are kept in hard-surfaced confinement systems. These cows often show signs of lameness at a young age and also become weak in the pasterns and coarse and stiff in the hocks. Animals whose rear legs are too straight and move poorly because of this condition should receive severe discrimination in the show ring.

The rear legs can also be too close at the hock when the individual walks and stands. Here, the weight is thrown at a bad angle on the hock and pastern joint, thus putting undue strain on them. The condition also causes the hooves to wear unevenly, which in turn worsens the condition and often causes lameness or awkward movement of the rear legs. Since this condition usually causes the rear feet to point outward, this deviation can be observed from either the side or the rear. The rear, however, provides the best view for a proper evaluation (Figs. 113 and 114). If severe, this is a serious discrimination.

A sickle-shaped rear leg detracts considerably from the straightness of the lines of a cow and is easily seen from the side view. From a functional standpoint, a slight to moderate degree of sickling of the rear leg is of little consequence. It rarely causes stiffness or lameness, and it rarely impairs the movement of the animal. In severe cases, however, it should be severely penalized because the weight is placed too far back on the heel and will set the stage for foot trouble. These cows often stand with their rear legs behind them instead of squarely beneath the rear quarters.

The bones in the leg can be light or weak (Fig. 104). This condition should be penalized only in very severe cases, for research has shown light bone to have a high positive association with herd life and lifetime production. Conversely, a coarse, rough bone is usually incompatible with good dairy character (Figs. 111 and 117). The discrimination may be slight to severe, depending on the degree of deviation from the ideal.

Front Legs

Deviations from the ideal front legs and the degree of discrimination are: enlarged knees (slight); bucked knees, knees bucked forward (serious); knees set back (slight); knock knees, close at the knees and toes out (slight to serious) (Fig. 126); bowlegged, legs bowed out and toes pointed in (slight to serious); crooked legs with toes pointing out (moderate) (Fig. 125); weak, too refined bone (slight to serious); legs too close together, causing insufficient chest room (slight to serious); poor feet and pasterns, as described for the rear feet and legs earlier in this chapter. Loosely attached or wing shoulders are often reflected in the front legs and can usually be detected by awkward movement and set of the front legs.

Crampiness

The feet and legs should be observed while the animal is walking and while standing. A hobbling, jerky, or awkward gait indicates flaws in conformation that must be evaluated to determine the proper discrimination. For example, arthritis or crampiness is given a serious discrimination both for cows and for bulls because it has such a marked influence on usefulness.

The general proportions of the legs must be considered. Every judge has observed legs that were too long and gangly, giving the effect of stilts, as well as too short or stubby legs, carrying the animal too close to the ground and making it appear heavy and compact in body conformation.

BREED AND SEX DIFFERENCES

Breed differences exist in feet and legs. Brown Swiss indisputably have the best feet and legs, and their feet seldom need trimming if the animals are given a reasonable amount of exercise on firm ground. Since their legs are generally well shaped, their weight is properly distributed when standing and when walking (Fig. 107). Brown Swiss breeders strive to maintain this characteristic by carefully discriminating against deviations.

The smaller breeds have received less attention and emphasis on legs and feet because deviations cause less trouble with a lightweight animal than with a larger one. The varying degrees of discrimination on deviations from the ideal feet and legs should be recognized by the judge who officiates for a number of breeds. This is important in order to avoid too much discrimination or too much tolerance while judging different breeds.

The set of hind legs, especially in young animals, can be considerably strengthened by extensive exercise. Controlled exercise on a motor-driven mechanical exerciser has proved extremely beneficial. Judges should give consideration to the influence of exercise on strength of legs and set of hocks. Heifers and cows in herds that have a tendency toward sickle legs show this characteristic to a much greater extent in the spring when they go out to pasture than they do 3 or 4 months later after they have had the benefit of considerable exercise while grazing on pasture.

Bulls are assigned considerably more points and criticized even more for feet and legs on the score card than are cows. This is justified because a bull carries more weight and through inheritance may transmit either good or bad legs and feet to his progeny. Thus, it is logical to start an improvement program with bulls, since they have so much more influence individually in a breeding program.

The straight hock, or post-legged condition, is serious in service bulls because the leg may buck forward in the hock during service. This may produce serious lameness and limit his usefulness. In extreme cases the hock joints become sore and inflamed (Fig. 131). A large, heavy bull may respond to this condition by shifting his weight from one side to another. As the condition becomes more pronounced, he increases the frequency of weight shifting, until the motion develops into a rhythm much like that of a bicycle rider. After a few weeks of this, the bull is usually in such poor condition that he can no longer stand or move about.

Evidence of arthritis or crampiness in the hind legs is given a serious discrimination, especially in bulls because it usually impairs their usefulness for service. This condition can be observed while the animal is walking and sometimes while standing, but particularly when in motion, since he is likely to shake the foot and leg periodically.

USEFUL TERMINOLOGY

Terminology used in describing various conditions of the feet and legs follows:

Legs:
1. Squarely placed legs to ensure good moving ability.
2. Legs are well placed; set squarely on the corners of the body.
3. Stands squarely on good legs (sound legs and feet).
4. Correct in set of legs.
5. Nearly perfect legs and feet.
6. Smooth (stylish) and nicely balanced set of legs and feet.
7. Ample bone (ample substance of bone).
8. Excellent set of legs, with ample, clean bone.
9. Sufficient substance and refinement of bone.
10. The gentle curve indicates a nearly perfect set to her legs.
11. Nearly perfect set of legs when viewed from the side and the rear, and while moving as well as standing.
12. Refined but sturdy legs.
13. Strong-boned but no sign of coarseness (clean-cut bone).
14. Small-boned but strong leg.
15. Good support below the hock.
16. Moves easily and strongly (a strong, free-moving leg).
17. Sickle-shaped hock and leg.
18. Easy on the hind legs (needs more support below the hock).
19. Entirely too much set to the hind leg.
20. Crooked legs, markedly apparent both on the move and while standing.
21. Too much set to the hock.
22. Post-legged (too straight on her legs).
23. Bone too coarse (thick) (heavy).
24. Bone too light (frail) (small) (refined).

25. Crippled legs (lacking in soundness of feet and legs).
26. Permanent (temporary) lameness due to leg (joint) injuries.
27. Too upstanding (too far off the ground) (too short a leg).
28. Too long and gangly in the legs (a bit too much length of legs).
29. Awkward set of the legs places too much pressure on the hock and pastern.

Hock:
1. Well-molded, clean, refined, but strong hock (clean at the hocks).
2. Well-placed hocks (points of hocks wide apart, allowing ample room for a wide rear udder).
3. Correct width and shape at the hocks.
4. Large, clean-cut, wide, and deep hocks.
5. Oversized (thick) (coarse) hock.
6. Meaty (puffy) condition of the hock.
7. Somewhat full and boggy in the hocks.
8. Set too close (wide) at the hocks.
9. Stands and walks so close at the hocks that the udder is pushed forward.
10. Slightly (moderately) (very pronounced) sickle-hocked legs.
11. Swollen (bruised) (inflamed) hock.

Pasterns:
1. Strong, smooth, well-proportioned, moderately long pasterns.
2. Nearly ideal set to her pasterns.
3. Strong pasterns that join the hoof smoothly.
4. She (he) possesses a common weakness of hind legs—weak pasterns.
5. Inflamed, sore (enlarged) pasterns.
6. Long, weak pasterns.
7. Ill-shaped pasterns, lacking in strength.
8. Too springy (soft) in the pasterns.
9. Long, weak pasterns that terminate in a weak foot.
10. Pasterns are too short and straight.
11. Too steep in the pasterns.

Feet:
1. Broad, well-shaped, deep foot, with solid walls and good wearing surface.
2. Great depth, width, and size of foot.
3. Deep, well-balanced foot, with proper depth at the heel.

4. Shapely feet with good wearing qualities.
5. Has a steep angle to her toe.
6. Has a short toe.
7. Possesses the preferred depth and width of heel.
8. Carries too much weight on the soft structure of the heel.
9. Feet seem very tender and she walks insecurely.
10. Walks too far back on her heels.
11. Shallow at the heels.
12. Infection and inflammation at the heel (on the bottom) (between the toes) of the foot.
13. Weight-bearing surface of the foot is too tender (too small) (very uneven) (too spongy).
14. Small feet with narrow and pointed toes.
15. Spreading toes (spraddle-toed) (foot opens up too much from the front).
16. Scar tissue between the toes is a sign of previous trouble (of foot trouble).
17. Shallow, flat foot that predisposes the individual to foot rot.
18. Swelling and lameness in the right (left) front (rear) foot.
19. Foot rot (infection) which has penetrated into the deeper tissues or joints.
20. Toes have grown too long.
21. Walls of the feet have curled under.
22. Essentially good feet but sorely in need of attention (trimming).
23. Badly neglected feet (lack of foot care) overemphasize the leg faults.
24. Temporary (permanent) lameness in her feet.

Front Legs:
1. Straight and well placed.
2. Should have greater width between the forelegs.
3. Stands and walks too close (not enough space between the front legs).
4. Knees too close (knock-kneed).
5. Legs curve outward (bow) too much.
6. Toes point outward.
7. Walks pigeon-toed (toes point inward) (paddling gait).
8. Knees set too far back on front legs.
9. Bucked knees (knees buck forward).
10. Awkward movement and set of front legs.
11. Swollen (bruised) (heavy) knee that lacks freedom of movement.

Figure 99 The strength of pastern and depth of foot in this cow are very good. (Courtesy American Guernsey Cattle Club, Peterborough, NH)

Figure 100 This cow has a shallow foot, and the hock is set too far back, which causes her to scuff when she walks and makes her susceptible to foot trouble (a moderate discrimination). Courtesy American Guernsey Cattle Club, Peterborough, NH)

Figure 101 A fine set of legs with strong pasterns and well-shaped feet. X-1 indicates point of thurl. The plumb line is even with the hock, X-2. The plumb should touch the ground midway between the heel and toe of the hoof. (Courtesy American Guernsey Cattle Club, Peterborough, NH)

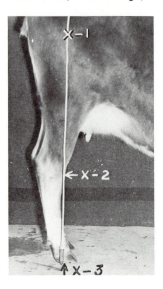

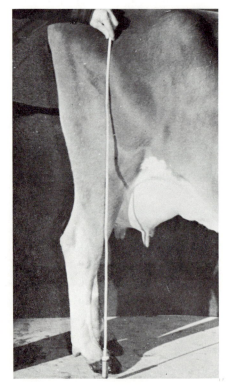

Figure 102 (Right) The plumb-line test, which can be duplicated in practice by an imaginary line, shows that the weak pasterns push the legs back too far. The space between the front of the hock and the plumb line is slight, but there would be none at all if this cow had a strong set of pasterns (a moderate discrimination). (Courtesy American Guernsey Cattle Club, Peterborough, NH)

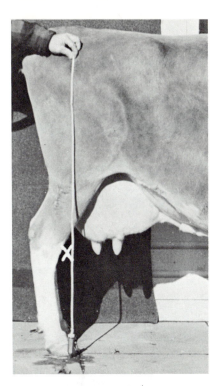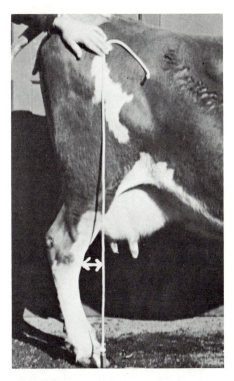

Figure 103 (Top Left) This cow's legs are sickle-shaped and hence awkward (a moderate discrimination). (Courtesy American Guernsey Cattle Club, Peterborough, NH)

Figure 104 This cow is very sickle-legged. Such legs are often referred to as willowy and are due to lack of substance in the bone (a serious discrimination). (Courtesy American Guernsey Cattle Club, Peterborough, NH)

Figure 105 Thick, coarse, and enlarged hock on a strong leg. This hock is easily injured and a constant source of trouble (a slight to moderate discrimination). (Courtesy American Guernsey Cattle Club, Peterborough, NH)

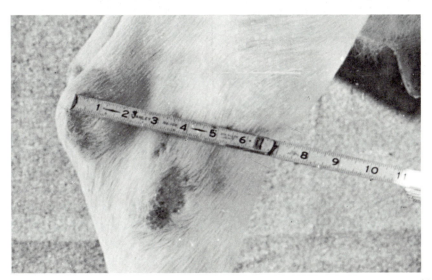

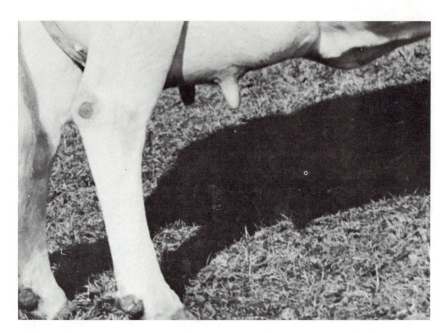

Figure 106   Ayrshire cow with strong leg bones and correct set of the hocks.

Figure 107   Brown Swiss cow having strong bones, ample refinement, and excellent set of the hocks. The legs are supported by strong pasterns, and the hoofs are deep and well rounded. (Courtesy Hycrest Farm, Leominster, MA)

Figure 108   These almost perfect legs on a Jersey cow have much more refinement of bone than those of the Brown Swiss in Fig. 107, but they are suitable for a smaller cow carrying considerably less weight.

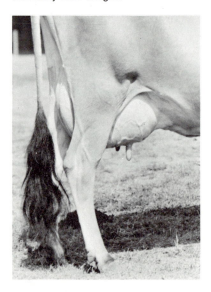

Figure 109 Almost ideal set of legs on a Guernsey cow. A splendid combination of correct proportions; clean-cut lines; refinement; strong, flat bones; and correct set of the hocks.

Figure 110 Guernsey cow whose legs are set acceptably but are too light in bone structure (a very slight discrimination).

Figure 111 In contrast to the cow shown in Fig. 109, this Guernsey stands on awkward legs that are too coarse and heavy in bone structure (a moderate discrimination).

Figure 112   Rear view of legs show-ing ideal width at hocks. (Courtesy Collins-Crest, Perry, NY)

Figure 113   The cow shown here stands too close at the hock. This brings the entire leg into an awkward position and causes the toes to point outward. The udder has an excellent rear attachment but is forced forward by the position of the rear legs (a moderate discrimination). (Courtesy American Guernsey Cattle Club, Peter-borough, NH)

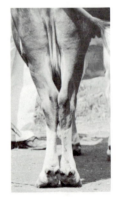

Figure 114   Cow showing the same conformation described in Fig. 113, but the condition here (close hocks, awkward legs, and feet that toe out) is much more pronounced (a serious discrimination).

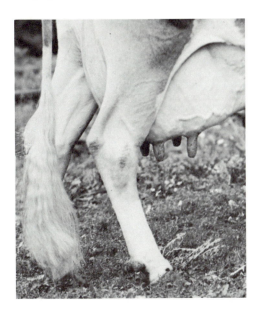

Figure 115 Legs that have too much curve in the hock and feet that point out at the toes. The curved hocks place the legs in an awkward position and support the weight at the wrong angle in this Jersey cow (a moderate discrimination).

Figure 116 The hocks of this Holstein cow have a nearly perfect set and can support the weight of the body with ease.

Figure 117 This sickle-legged Ayrshire cow shows coarse bones and faulty pasterns (a serious discrimination).

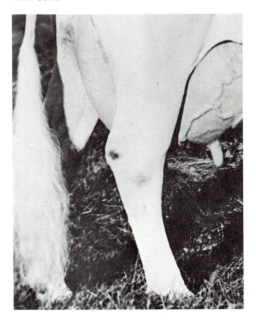

Figure 118 The hocks of this Guernsey cow are acceptable, but the pasterns are extremely weak, and the feet are too long and shallow (a moderate to serious discrimination).

Figure 119 Sickle legs, weak pasterns, shallow feet, and low hoof angle cause this Holstein cow to walk on the tender part of the heel of her foot. This condition is conducive to foot trouble and usually shortens the useful life (a serious discrimination).

Figure 120 Extremely sickle legs, long weak pasterns, and low hoof angle on a high-producing Guernsey cow (a serious discrimination).

Figure 121 The hocks of this Ayrshire cow are too straight, and the pasterns are broken and deformed (a serious discrimination). The strain caused by this condition will shorten the useful and productive life of a cow.

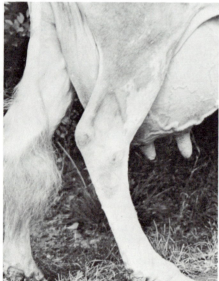

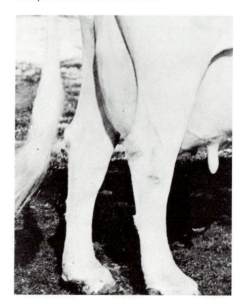

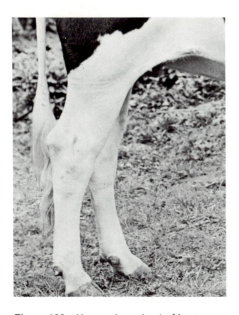

Figure 122 Very awkward set of legs on a Holstein heifer.

Figure 123 Ayrshire heifer whose legs are very ungainly. Time will only exaggerate this condition.

Figure 124 Almost ideal front legs of a Guernsey cow. Note the deep and well-shaped foot, strong pasterns, straight leg, width at well-formed knees, and ample chest room. (Courtesy American Guernsey Cattle Club, Peterborough, NH)

Figure 125 Crooked front legs and feet that toe out. A twist in the legs throws the toes out and places the weight on the feet at an abnormal angle (a moderate discrimination). (Courtesy American Guernsey Cattle Club, Peterborough, NH)

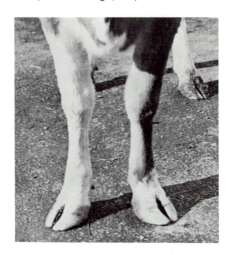

Figure 126 Poor front legs with knock-knees and feet toeing out. The legs are so close together that insufficient space is allowed for the chest area (a serious discrimination). (Courtesy American Guernsey Cattle Club, Peterborough, NH)

Figure 127 Almost perfect legs of an Ayrshire bull. Note the combination of strength, refinement, and clean mold of bone. The set of the hock is ideal, the pasterns are firm, strong, of proper length, and are attached to wide, deep, and well-formed feet. The fine shape of these feet and legs was maintained up to the death of this bull at nearly 15 years of age. (Courtesy Curtiss Candy Company Farm, Cary, IL)

Figure 128 Strong, refined legs of a Guernsey bull. Note the flatness and strength of bone in the leg and the clean mold to the hock with the correct set. (Courtesy McDonald Farms, Cortland, NY)

Figure 129 Awkward legs of a young Guernsey bull. A great deal of this awkwardness is due to long, weak pasterns, shallow heel, and low hoof angle (a severe discrimination).

Figure 131 Faulty legs, which are too straight in the hock and too weak in the pasterns and feet for a large mature bull. Good rear legs are necessary to support the weight properly during service. Legs such as these handicap the bull and usually shorten his useful life (a serious discrimination).

Figure 130 Poor legs of a Jersey bull. The heels are shallow, and the pasterns are weak. This is a serious discrimination in a bull. The weakness was transmitted to a majority of his sons and daughters.

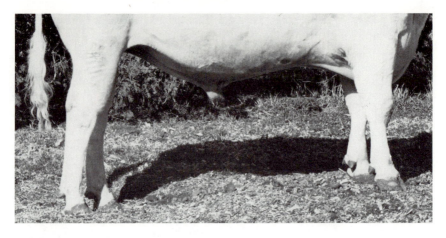

Figure 132   The legs on this bull are far too straight in the hock, and the front feet toe out (a serious discrimination).

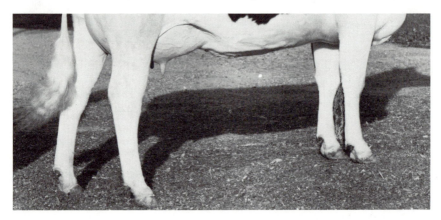

Figure 133   An outstanding young cow except that the hock is too straight and the pastern is steep and awkward (a serious discrimination).

Figure 134   A good Guernsey heifer who is sickle-hocked and who carries her legs too far back (a moderate discrimination). She was Reserve All-American.

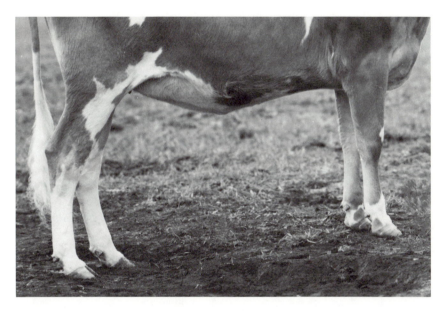

Figure 135 A Guernsey heifer who has broken pasterns, low hoof angle, and shallow heel (a serious discrimination).

Figure 136 Side and rear view of a foot with trouble caused by a poorly shaped and shallow hoof.

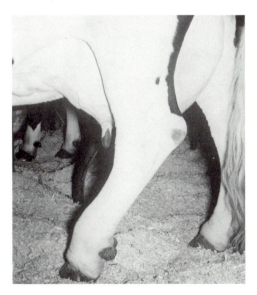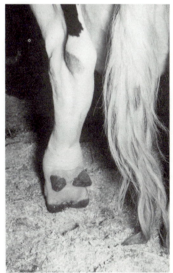

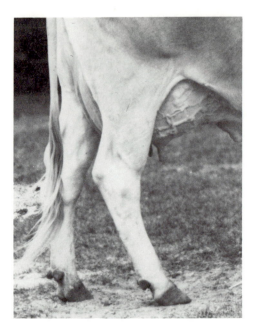

Figure 137   A 3-year-old Brown Swiss with sickle legs. Rarely do individuals of this breed have sickle legs, and seldom do they get the top spot in a show if they do.

Figure 138   This Ex 97, Reserve All-Time All-American, three times All-American and two times Reserve Bull, was a many-time winner at state fairs and national shows. His nearly perfect legs and feet literally permitted him to trot around the show ring at almost 9 years of age. (Courtesy C. M. Bottema, Jr., and son, Indianapolis, IN)

Figure 139 This Ex 96 Holstein cow was designated All-Time All-American 4-year-old. She also was an All-American and Reserve All-American Aged Cow. This picture, taken when she was 5 years old, shows an almost perfect set of legs and feet. Note the strong, flat bone, the clean hock, the general shape of the hocks, and the deep, well-formed foot. (Courtesy Hanover Hill, Armenia, NY)

Figure 140 The depth of foot, especially the depth of heel, is superior on this outstanding set of feet and legs on this high-producing 3E 93 cow. (Courtesy Carnation Farms, Carnation, WA)

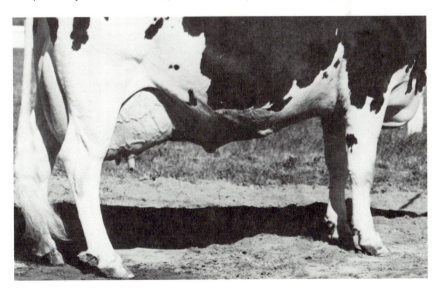

9
BODY CAPACITY

A large body is very important in a cow because it promotes life-giving functions. The most efficient and economical production is obtained from forage. The more high-quality forages a cow can consume, the less grain is required in the ration to supply the nutrients for production. The trend today is toward feeding moderate amounts of grain for efficient production and toward providing high-quality and easily digestible forage. This serves the purpose of the ruminant (complex stomach versus simple) and makes the dairy cow less dependent on the food supply that can be used for human consumption.

The type of body dictates the kind of ration that can be handled efficiently, and the kind of ration and feeding management have a distinct influence on efficient, economical production as well as on body type in dairy cattle.

PREFERRED TYPE OF BODY

A long, well-sprung rib gives a capacious middle or barrel capacity (Fig. 141). The ribs should be long, deep, wide, strong, and arched. Body capacity is controlled by length, depth, and spring or degree of arch of rib. Since length, depth, and width all exert a great influence on total body capacity, they must be carefully evaluated.

In more recent years, there has been less emphasis on the importance of body capacity in dairy cattle. The PDCA has reduced the point value on its score card from 20 to 10, but it still recognizes its importance by approving a relatively large body in proportion to the size, age, and stage of gestation of the animal, a body that provides ample capacity, strength, and vigor. A large body means one that has a large heart girth, formed by long, well-sprung fore ribs, showing ample fullness in the crops and at the point of elbow in the region behind the forearm. This results in a moderately deep and wide chest floor with ample strength of heart. The barrel should be long and strongly supported, with well-sprung, widely spaced ribs, and depth and width should increase toward the rear of the barrel. This should not, however, conflict with dairy quality (refinement with strength) and there should be no sign of coarseness.

DEVIATIONS FROM THE IDEAL

After carefully studying the description of ideal body conformation, one should see that the possible deviations are obvious. From the front view, a cow can be very narrow and pinched in heart girth, thus having insufficient room for the heart and lungs. Moderate heart girth is important for a high-producing cow. A cow that is too narrow in this region is also usually frail in general conformation. Cows that are too heavy and coarse in this region lack dairy quality. The latter is a more serious discrimination than the former, but either deviation, if marked, is a serious discrimination.

From the side view it is possible to observe many deficiencies in body. These range from serious defects such as a very short body (a serious discrimination) to a slight defect in fullness of heart girth or depth of ribs (a slight discrimination). Sometimes this deficiency in depth of ribbing, if severe, is exhibited in both fore and rear rib and makes the body much too shallow (a serious discrimination). Depth of rear flank is also observed from the side, but its practical significance is not so great as that for depth of ribs. Therefore, it receives less emphasis; it does, however, contribute to the general appearance of the cow. It is also very important that the individual rib bones are flat and wide apart because they give openness of body and sufficient length. A compact body with the ribs close together is a serious discrimination in a dairy animal, especially in mature cows.

Spring of ribs is best studied by standing directly behind the cow (Figs. 160 and 161). Any lack of fullness behind the shoulders in the region of the crops and any lack of spring of fore and rear ribs can be determined from this position. A body too flat because of a lack of spring of rib has a pronounced effect on capacity and must be evaluated from this standpoint.

FACTORS AFFECTING BODY TYPE

The influence of sex, pregnancy, stage of lactation, level of production, age, sickness, and many other factors must be given consideration if one is to evaluate body capacity properly.

Sex is taken into consideration on the score card, which assigns 10 points to the body capacity of a cow and 15 points to the body capacity of a bull.

Stage of pregnancy has considerable influence on body shape and, as mentioned above, should receive due consideration. During the last 3 or 4 months of pregnancy a cow, if fed properly, takes on weight, and this should be considered for proper evaluation, especially when the cow is being judged on the condition of fleshing and openness of ribbing. Degree of fleshing is also closely associated with stage of lactation. At the peak of her production, a good milker should be open and lean of body. At this stage she may be milked down so that she is not so full in heart girth or so deep through the body. Stage of lactation should be taken into account, and if the cow appears to be a hard worker displaying great dairy quality, due emphasis should be given to these characteristics, with little discrimination for a temporary lack in extreme depth and width of body, especially in the chest area. A narrow chest should not be criticized if due to dairy quality.

Age has a considerable influence on body shape. Young cows have firmer muscle tone than do old cows, and they have considerably less girth and depth of barrel. This is effectively demonstrated in Figures 2 through 5, which show the body shape of Jane of Vernon at the ages of 3, 4, 11, and 15. Heifers and young cows develop more body capacity as they advance in age; therefore, a slight deficiency, especially in depth of body, can often be overlooked in heifers, especially in the junior yearling stage in which growth is rapid and in recently fresh 2- and 3-year-old cows.

Sickness frequently affects the degree of fill a cow needs to show her body off to best advantage, but this does not deceive a good judge. Actually, when a good cow is temporarily lacking in fill, one can best appreciate her fine depth of body and length of rib. In contrast to this, some cows that have poor body development are finicky eaters and frequently go off feed. If sickness is given as an excuse for lack of body and fill, the experienced judge is not deceived.

A tight-ribbed heifer or cow can be changed in appearance by feeding her large quantities of forage so that she appears acceptable, but again the experienced judge will not be fooled. Animals that have proper depth can be conditioned to capitalize fully on this good conformation by feeding on a ration consisting predominantly of forage or beet pulp.

DESCRIPTIVE TERMINOLOGY

To help the student develop good reasons for placing animals in the show ring, the following descriptive terms are listed for the ideal and for deviations from the ideal:

1. Powerful-bodied cow with tremendous length, depth, openness, and spring of rib.
2. Beautifully arched (sprung) ribs, combined with ample length and depth,

give this individual nearly ideal body (barrel) (feed-handling) capacity.

3. Wide, refined chest, deep heart and barrel, plus exceptional spring of rib give her a most outstanding body capacity that must be considered among the best of the breed.

4. Long, well-sprung ribs give her a capacious middle and barrel.

5. Although a trifle flat and deficient in the crops and pinched in the heart region immediately behind the forearm (point of elbow), she is a large cow with very good spring and depth of rear rib.

6. Fuller in the crops because of more spring of fore rib.

7. Deeper heart and fore rib.

8. Wider (broader) chest (floor of chest).

9. Stronger (greater strength or fuller back of the fore arm).

10. Greater (wider) spring of rib.

11. More arch to her ribs both fore and rear.

12. Deeper flanked cow.

13. She displays a longer, deeper, wider, more capacious barrel.

14. Ample width and strength of chest, but too thick, coarse, and meaty in the brisket.

15. Too frail, overrefined, and weak in the chest.

16. Weak, narrow chest.

17. Short, closely spaced ribs give her a round-ribbed appearance, greatly lacking in dairy quality.

18. Pinched in the heart.

19. Lacks depth and fullness of heart (barrel) (rear flank).

20. Short and pinched in the fore rib (heart) (barrel).

21. Cuts in behind the shoulder and lacks spring in fore rib.

22. Very untidy body that sags in the middle (barrel).

23. Potbellied conformation.

24. Narrow, slab-sided, flat-bodied individual.

25. Great spring of rib and excellent from the rear view, but a round, shallow body that lacks openness of rib when viewed from the side.

26. Greatly lacking in depth, openness of rib, and length of body.

27. Great width of loin and body.

28. A large, capacious barrel.

The student should study Figures 141 through 161 carefully in order to develop powers of observation and to form definite impressions, which are so important in precision judging. The experienced judge would see differences in these pictures at a glance, but the beginner should study and observe the animals until he or she can visualize quickly the contrasting differences.

Figure 141 Large, mature Holstein cow with long, strong, deep body; sufficient dairy quality; and openness of rib. In mature form (Ex 96-2E) she produced 31,028 lb of milk and 1020 lb of fat at 5 years of age. She was Reserve All-American 4-Year-Old the previous year. She has a production pedigree and made over 20,000 lb of milk in each of her first four lactations. Although she is big (60 inches tall), she is a refined kind of cow with excellent classification in every breakdown. (Courtesy Collins-Crest Farm, Perry, NY)

Figure 142 A Jersey cow having outstanding body capacity. She has exceptional length of body, fullness of heart, and openness of rib with adequate depth. This outstanding Ex 97 Jersey cow was sired by a bull having a high-predicted difference for milk production. She has produced up to 25,010 lb of milk and 1119 lb of fat. She had three consecutive records well over 20,000 lb of milk at 4 years, 1 month; at 5 years, 2 months; and at 6 years, 3 months of age. She was Grand Champion and Best-Uddered Aged Cow at the National Jersey Show and later was designated All-American Aged Cow. (Courtesy Happy Valley Farm, Danville, KY, and Briggs and Beth Cunningham, Newberry, SC)

Figure 143 For a mature cow, this individual shows a deficiency in body. She is tight and restricted in the heart and is not particularly deep in the barrel region. (Courtesy Cornell University, Ithaca, NY)

Figure 144 The cow pictured here is mediocre, as compared with the one in Fig. 141. She has an acceptable body but only medium depth.

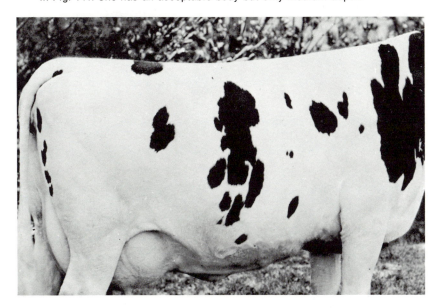

Figure 145 To provide judgment on depth of body, a Holstein cow that classified 93-3E with excellent across the board and a production of 25,852 lb of milk, 3.9%, 1001 lb of fat at 5 years is pictured at two ages. The more developed stage is pictured above, and the less developed stage is below.

Figure 146 The same cow as in Fig. 145 but at a younger stage with less development. The beginner may not observe the differences immediately, but the experienced judge sees the contrast at once.

Figure 147 The right kind of body for a young cow may be observed in this Reserve All-American 2-year-old. (Courtesy Woodacres Farm, Princeton, NJ)

Figure 148 A daughter of the cow in Fig. 147 displays the same preferred depth of body. Note the depth and fullness of heart. (Courtesy Woodacres Farm, Princeton, NJ)

Figure 149   This All-American Ex 96 Holstein cow, pictured here at 3 years of age, will be presented at 2-year intervals to 9 years of age. These pictures will provide an opportunity to study changes with age. (Courtesy Hanover Hill, Armenia, NY)

Figure 150   The same cow as in Fig. 149 at 5 years of age. At 4 years she was designated All-Time All-American. She was nominated for All-American 6 consecutive years. (Courtesy Hanover Hill, Armenia, NY)

Figure 151 In bloom at 7 years of age, the young cow pictured in Fig. 149 now displays great depth, length, and width (spring of rib) of a well-supported and proportioned body. She was All-American Aged Cow at 7 years of age. (Courtesy Hanover Hill, Armenia, NY)

Figure 152 At 9 years of age this great cow is showing a slight change in muscle tone, but her udder, legs and feet, and general conformation are a real credit to her previous winnings in the show ring. (Courtesy Hanover Hill, Armenia, NY)

Figure 153 This cow shows a deficiency in capacity and depth of body, especially in the rear ribs.

Figure 154 A conspicuous deficiency in body capacity is indicated here by a general lack of strength and capacity, especially in the short, shallow fore and rear ribs.

Figure 155 The deep heart and fore and rear ribs of this 4E 97 Holstein cow provide a superior body capacity. (Courtesy Bob and Kay Miller and family, Mil-R-Mor, Dundee, IL)

Figure 156 This Brown Swiss cow has deep fore and rear ribs but lacks length of body and fullness behind the forearm and in the crops.

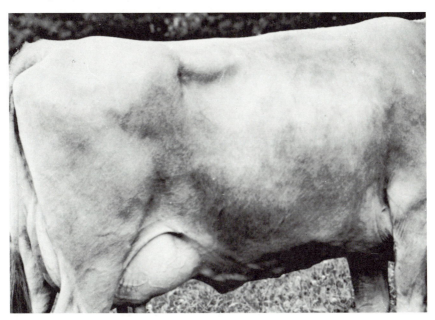

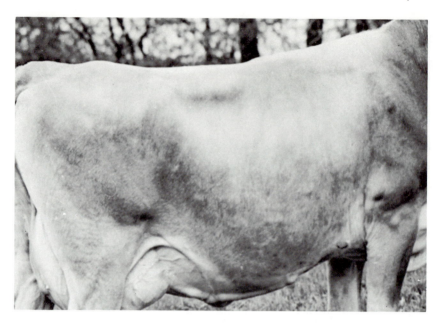

Figure 157   Brown Swiss cow having a deficiency in heart girth and a short body.

Figure 158   A short body, somewhat lacking in depth of heart, is the flaw in this Brown Swiss cow. Compare Figs. 155 through 158 which are arranged in descending order of body capacity.

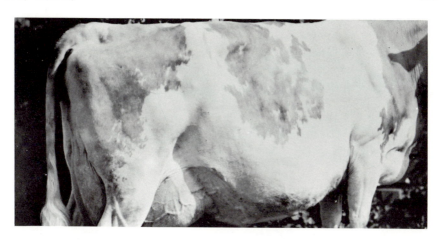

Figure 159 This Guernsey cow must be criticized for being too large and untidy in body conformation, even though the photograph was taken when she was close to calving. Cows having such conformation usually have a tremendous feed-handling capacity, but they are awkward and more prone to injuries.

Figure 160 Excellent spring of rib can be observed from the back view of this Holstein cow.

Figure 161 This cow, in contrast to the one pictured in Fig. 160, is extremely deficient in spring of fore rib, as indicated by a noticeable insufficiency in the region of the crops. She is also narrow over the loin and rump.

10

TOPLINES

The topline of cows and bulls probably does not have a great influence on productive capacity, but in many ways it reflects the general strength and conformation of the individual and adds greatly to overall balance and symmetry.

General muscle tone, the health of the animal, the condition of the legs and feet, and other characteristics of refinement and strength are indicated by the topline. Openness of form throughout the back and loin region is one of the major indications of dairy quality. Strength of frame development can also be observed here. A strong, straight back may be helpful to the mammary, reproductive, and digestive systems. The vertebral column, which forms the back, carries the spinal cord; thus, marked irregularities in the spine can affect the disposition and wearing qualities, causing premature aging in some individuals.

Proper rump conformation is important because the rump provides the support and roof for the udder. It is possible, however, for a cow to have a well-formed rump but poor udder attachments. Strength of loin and a square, nearly level, wide rump can affect the entire skeletal structure in the rear quarters. This, in turn, can determine ease of calving which, together with the way the reproductive organs are carried, greatly influences reproductive efficiency. This applies not only to individuals but also to groups, for example, the daughters of a particular sire. It is now generally recognized that a rump with the pin bones slightly lower than the hips is preferred because it results in better uterine drainage and improved health of the genital tract.

PREFERRED TYPE FOR TOPLINES

The entire topline from the withers to the tail head, comprising the back and rump, should form a nearly straight line, with the pins slightly lower than the hips (Fig. 162). The back, made up of the chine and loin, should be strong and straight, with well-defined vertebrae. The loin should be broad, strong, and nearly level.

A long, wide, nearly level pelvic region or rump is ideal. Important parts of the rump are the hips, thurls, pin bones, and tail head. The hip points, or hooks, should be prominently displayed and there should be ample width between them. They should be clean (free from excess flesh) and they should be approximately level laterally with the back. Prominent hips are associated with dairy character.

The thurls on a good rump should be wide apart and high enough so that they add to the general fullness and levelness of the pelvic region. They should be centrally located between the hips and pins. Thurls located too far back are often associated with faulty leg structure. In addition to being wide apart, the pin bones should be well defined and free from fatty or excess tissue deposits. The region from hips to pins should be long and nearly level, with the pins set slightly lower than the hips, especially in young animals.

The tail head should show refinement and should form a smooth, level ending of a straight topline. The tail head should be slightly above, and neatly set between, the pin bones. Some breed differences are tolerated. For example, a high, somewhat prominent tail head receives only a slight discrimination in Brown Swiss but is penalized considerably more in other breeds. Jerseys, especially noted for their correct rumps, occasionally have wide, flat tail heads, but these are not objectionable if they are not patchy and covered with excess fatty tissue. In judging Ayrshires, Guernseys, and Brown Swiss, fullness of rump is not considered as important, especially for young animals, as for Jerseys and Holsteins.

DEVIATIONS FROM THE PREFERRED TYPE

Each region described for the preferred type of topline has various deviations which carry different degrees of discrimination, depending on how pronounced the objectionable characteristic is.

The chine region is sometimes heavy and coarse, with the vertebrae placed close together. This condition is very objectionable. The same is true for a noticeable sag behind the withers or at the junction of the chine and loin. This sagging indicates weakness. An arch in this region, produced by several vertebrae having been pushed up too high, is given only a slight discrimination unless the conditon is so severe that it may eventually have a crippling effect.

Many deviations and degrees of deviation exist in the loin region. A narrow loin is a sign of weakness and is usually associated with a narrow body. It should receive a moderate to heavy discrimination. A severe dip in the loin is serious. A heavy, rounded loin, lacking in prominent vertebrae and covered with fatty tissue, is given a serious discrimination because it indicates a lack of dairy character and refinement.

Hips that are very rounded and narrow are objectionable because they exemplify compactness and lack of dairy quality. A capped hip, caused by a knocked-down point of hip, is not a discrimination unless it affects the mobility of the animal. A heavy covering of excess tissue is very objectionable, especially during the time of lactation, when it should be milked off. The thurls should not be set too low or be so narrow that they give an A-shaped appearance to the area between the hips and pins. This condition, which constricts the pelvic region, is assigned a moderate discrimination.

Pin bones that are pinched or set close together are often conducive to difficult parturition. Pin bones that are patchy and covered with excess fatty tissue, indicating a lack of productive ability or improper feeding, receive serious discriminations.

Objectionable features of the tail head include lack of smoothness, abrupt rounding off, too much height and prominence, deeply recessed between the pins, and roughness and coarseness. A recessed tail head is often accompanied by a horizontal or partially horizontal vulva which is conducive to poor drainage and increased contamination of the genital tract. Both of these conditions may impair efficient reproductive performance. A high, rough, or coarse tail head may be caused by a cystic or diseased ovary and is also of practical or functional importance. Discrimination for all these deviations ranges from slight to serious; a serious penalty is scored only when the condition is severe.

Deviations from the proper rump conformation include the following: insufficient length due to short distance from hip to pins; lack of width due to insufficient distance between hips, thurls, and pins; low thurls, which cause the rump to "shed off" too abruptly on each side; and low pins, which result in a sloping rump. If both low thurls and low pins are present in the same animal, the rump may slope both sideways and toward the back. Since conformation in the pelvic region is closely associated with ease in calf-bearing and other reproductive efficiency factors, the discrimination is serious for a marked deficiency in the points listed for overall rump structure.

TERMINOLOGY FOR REASONS ON PLACINGS

The following terms are useful in describing reasons for ranking individuals on placings for toplines:

1. Strong, well-carried back.
2. Straight and strong in the back, with vertebrae well defined.
3. Level through the region of the chine and loin.

4. Loin that is wide (broad), flat (smooth), strong, long, and free of excess flesh.
5. Narrow-loined individual.
6. Weak back (a bit slack in the back) (low in the back).
7. Lazy in the back (sags) (dips) or loin.
8. "Sleepy" top (low in the topline) (settled in the top when standing).
9. Easy (weak) (sags) in the back.
10. Slight (severe) dip at the loin disrupts the straight line of the vertebrae.
11. Lacking in width and proper flatness of loin.
12. Arched chine and arches (bridges) (roaches) at the loin.
13. Wide, prominent, well-defined hips (hip bones high and wide apart).
14. Level, sharp hooks, free from excess tissue (flesh).
15. Broad, prominent hips that show a lot of dairy character.
16. Narrow, rounded hips, indicating a lack of dairy character.
17. Capped hip (point knocked off).
18. Hips that are much too patchy.
19. Thurls up high and wide apart to help form a square rump and provide roominess in the pelvic region (higher and wider at the thurls) (thurls high and broad) (comparatively full above the thurls).
20. Low, narrow thurls forming a rump that sheds off to the side.
21. Wide, smooth, well-defined, and nearly level pin bones (wide and high at the pins).
22. Narrow and pinched at the pins, accompanied by a narrow rear udder attachment.
23. Dropping off at the pins (low at the pins).
24. Tail head set in smoothly.
25. Refinement and clean-cutness of the level tail head.
26. Tail head drops away too abruptly (does not carry out far enough).
27. Recessed tail head.
28. Flat, patchy, coarse tail head.
29. High and coarse tail setting.
30. Slight discrimination for wry tail angled to the right (left).
31. Long (proper length) from hooks (hips) to pins.
32. Nearly level from hooks to pins.
33. Wide across the rump (broad, with roomy pelvis, nearly level laterally).
34. Carrying out well and nearly level over the pelvic region.
35. Square rump with proper width at the hooks, an advantage in wide, nearly level pins and well filled out in the thurls. Also neatly laid-in at the tail setting (head).

36. Too narrow in the rump, especially at the pins, which are very restricted.

37. Too much slope in the rump.

38. Droopy rump.

39. Low narrow thurls and pins.

40. Rump slopes from hips to pins.

41. Square rump, but coarse tail setting.

42. Rough, irregular rump, with a niche near the tail head.

43. Peaked over the rump.

44. High or coarse in the pelvic arch.

45. Extremely pleasing topline, indicating a strong back, and terminating in a square rump made up of wide, prominent hips; high, wide thurls; nearly level from hips to pins; and great width and length over the entire pelvic region.

46. Very wavy and irregular top, with a drop in the chine, an arch in the loin, and tilt to the rump caused by low pins and thurls.

47. Arched chine, dipped loin, and rough irregular rump that sheds off to the side and back and carries a high, rough tail head.

48. Straight, level, nearly perfect topline, with open, well-defined, well-spaced vertebrae, but a slight discrimination must be made for the broken tail head.

Most good and bad points of conformation discussed in this chapter are illustrated in varying degrees in Figure 162 through 189. Some of these are arranged in descending order and some by contrast to make them more useful from a teaching standpoint and to show the relative importance of the deviations that are illustrated and described.

Figure 162   This nearly perfect topline is on a 4E 97-point high-producing cow. Note the high, sharp withers; strong, smooth chine; wide, level loin; and wide, nearly level rump with pin bones in the preferred position, slightly lower than the hips (photo reversed). (Courtesy Bob and Kay Miller and family, Mil-R-Mor, Dundee, IL)

Figure 163   A very straight and level topline, but it is not as strong in the loin, and is a trifle higher at the point of withers than the animal in Fig. 162. These slight differences do not carry much practical significance, but they must be considered in precision judging.

Figure 164   The topline of the famous Jane of Vernon is very satisfactory. The wide, flat, well-defined loin shows a very slight dip, but the top terminates in a long, nearly level rump. This softness in the loin became pronounced at 16 years of age. (Courtesy Brown Swiss Breeders Association, Beloit, WI)

Figure 165   Topline very similar to that shown in Fig. 164, but the loin does not show quite as much width, does not have the preferred flatness, and the pin bones are a bit too high (a slight discrimination).

Figure 166   This top shows a slight rise at the withers and a very slight sag in the back (a moderate discrimination).

Figure 167   An acceptable and satisfactory top, but it shows a slight unevenness in the back and is not as flat or smooth over the rump as is preferred (a slight discrimination).

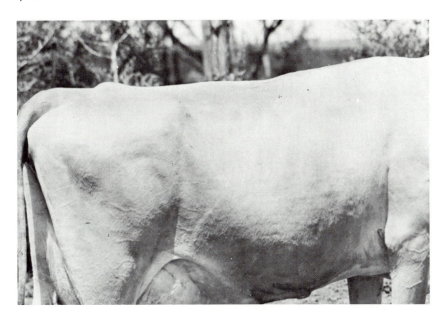

Figure 168   Satisfactory top, although a trifle wavy in the back, with a slight niche on top of the rump and a high pelvic arch (a slight discrimination).

Figure 169   Topline similar to that shown in Fig. 168, but the irregularity of the back is slightly more pronounced, with the hooks higher and the pins lower; the rump is niched and the thurl region lacks fullness (a slight to moderate discrimination).

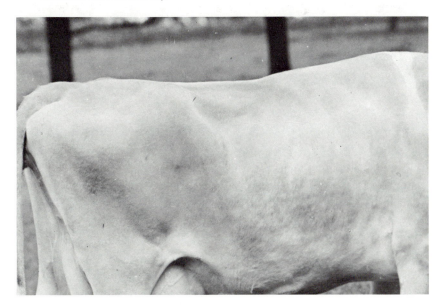

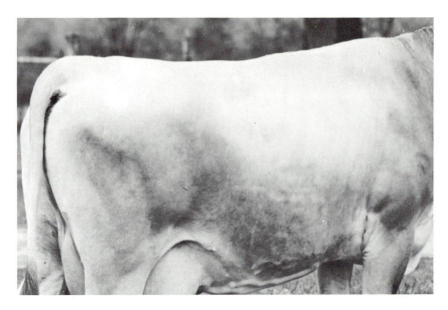

Figure 170   This top shows a slight wave in the back, low pins, low thurls, and a rough tail head (a moderate discrimination).

Figure 171   This topline is wavy, especially over the pelvic region, and drops away at the rump because the pins and thurls are too low (a moderate discrimination).

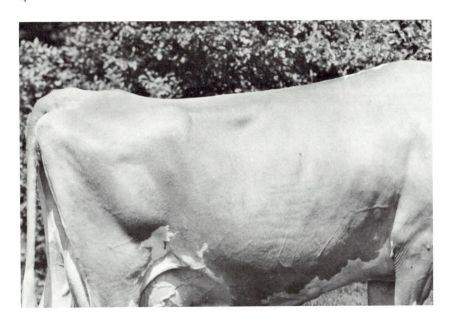

Figure 172 Low loin and slight sag in the back; pins are slightly higher than hips, thurls are low, and the cow is rough over the pelvic region (a moderate to serious discrimination).

Figure 173 Good back but droopy rump (moderate to serious discrimination).

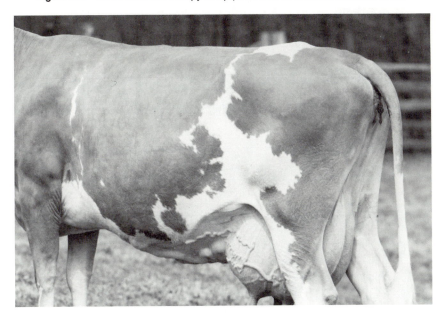

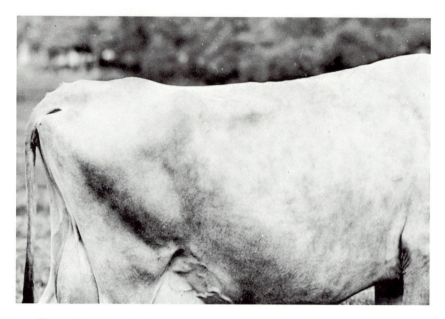

Figure 174 Acceptable back, but a very rough rump that is low at the pins and thurls and high and coarse in the pelvic arch (a moderate to serious discrimination).

Figure 175 Serious discrimination should be given to this poor topline. Note dipped, narrow loin; rough pelvic region; and narrow rump that sheds off very abruptly on the side and is low at the pins.

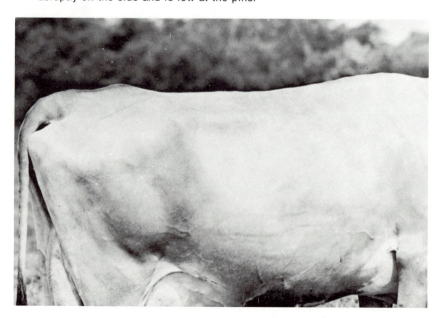

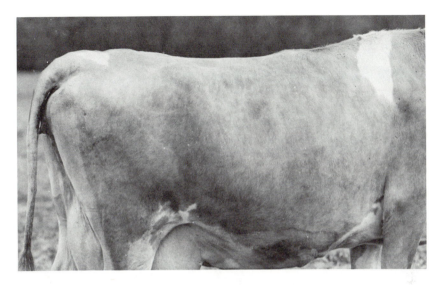

Figure 176 Rough back, with a drop at the end of the chine and a pelvic region with low pins and thurls (a serious discrimination).

Figure 177 Severe dip in loin and pins higher than hips badly disrupt the straight line of the vertebrae of this young cow and may impair her reproductive efficiency (a serious discrimination). Usually the condition gets worse as a cow advances in age, and it prevents normal wear and aging and often leads to reproductive problems.

Figure 178   This very wavy top indicates a weakness in overall conformation that will undoubtedly affect wearing qualities (a serious discrimination).

Figure 179   This very weak back, especially in the region of the loin, and a rump that sheds off both sideways and backward must be considered serious deviations from the ideal type.

Figure 180 This rump is so steep in its slope that it should be given a serious discrimination. Note that the malformed structure of the pelvic region has caused the udder to be shoved forward and tilted.

Figure 181 Square, wide rump that appears nearly level and ideal from the rear view. (Courtesy Holstein-Friesian Association of America, Brattleboro, VT)

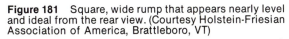

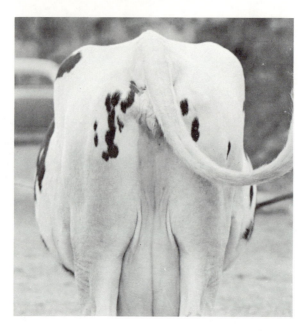

Figure 182 Droopy rump caused by low pins. (Courtesy Holstein-Friesian Association of America, Brattleboro, VT)

Figure 183 This young cow, unanimous 3-year-old All-American, is far advanced in general appearance, dairy character, udder shape and quality, and displays a near perfect topline. Her sharpness at the withers, clean and strong vertebrae throughout the back, wide hips, and thurls and pins which are slightly lower than the hips are all positive. Her straight top without a sign of patchiness or fat deposits all combine for a topline that can serve as the ideal. She produced over 15,000 lb M at 2 yr 1 mo and well over 20,000 at 3 yr 2 mo. (Courtesy Jack E. and Daria C. Stookey, Leesburg, IN)

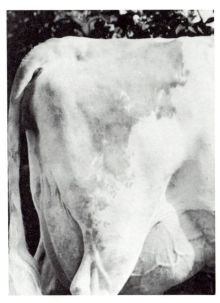

Figure 184 This rump slopes too abruptly at the sides because of low thurls. (Courtesy Holstein-Friesian Association of America, Brattleboro, VT)

Figure 185 This cow was photographed just before calving. The rump ligaments are relaxed and the tail head more prominent. This should be taken into consideration when evaluating a rump at this stage.

Figure 186 Strong, straight topline on this Ayrshire bull terminates a nearly level rump.

Figure 187   In contrast to the topline shown in Fig. 186, this one is very wavy and terminates in a rump that is considerably shorter from hips to pins; the pinched, closely set pins are the cause of a prominent, coarse tail head (a moderate discrimination).

Figure 188   An acceptable back, but the rump terminates in a high, prominent tail head (a moderate discrimination).

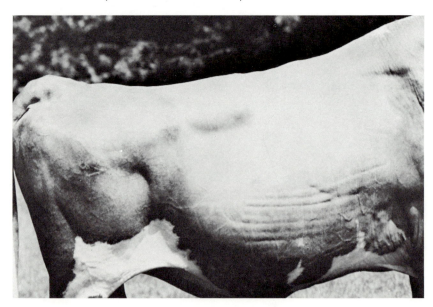

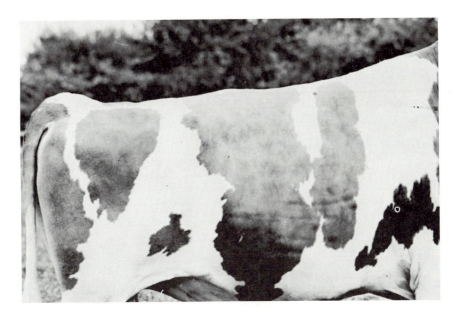

Figure 189 The topline on this Guernsey bull should receive more discrimination than that shown in Fig. 188 because the back arches in the region of the loin, the pins are pinched, and the tail head structure is even higher (a moderate to serious discrimination).

11

UDDER SHAPE AND ATTACHMENTS

The udder is the most useful and, therefore, the most important part of a cow. A good udder is so essential that it receives the greatest emphasis in judging. Since 35 of the total points, or emphases, are assigned to the udder in the score card, the animal's success in the show ring may well depend on the judge's evaluation of her udder.

Marked improvement in udders has been evident for all breeds over the past two decades, but was accelerated when the medial suspensory ligament was included in the functional type traits in the revision of the Holstein type classification system. Other breeds subsequently placed increased emphasis on this functional trait. This represents the results of judgment and selection according to a specific ideal, and it is an indication of what can be accomplished through breeding in dairy cattle.

Every practical dairy farmer knows that udders of poor conformation are more susceptible to mastitis and injury, which result in higher veterinary costs and reduced production. High production is one of the severest tests of, and strains on, an udder. Some high producers have udders far short of the ideal. If there is an inherited weakness in shape or attachments, the udder will not stand up under high production.

The mammary system, including rear udder, suspensory ligament, fore udder, and teats, has been found to have a positive association with milk production in Jerseys[1] and herd life and lifetime production in Holsteins. Broken suspensory ligament, loose or broken rear udder, and undesirably shaped teats have been shown to contribute to shorter herd life and lowered lifetime production.[2]

THE IDEAL UDDER

Every successful judge of dairy cattle must have a clear concept of the ideal udder if he or she is to evaluate properly deviations from the ideal and arrive at a correct placing.

The preferred type of udder is much alike for all breeds. An almost perfect udder for each breed is shown in Figures 25, 198 through 207 and 255. The rear udder should be high, wide, strong, and smooth in its attachment, and it should have moderate depth and large capacity (Fig. 229). It generally produces approximately 60 percent of the total milk. It should be gently rounded, with both quarters the same size in order to give it balance. The rear teats should be perpendicular, or at right angles, from the floor of the udder, and near the back corners, with a moderate distance between the two.

The floor of the ideal rear udder usually has a well-defined cleft, or division; the division (once inadvisedly criticized by some judges) is now considered very important because it reflects a strong medial suspensory ligament. An udder having a perfectly level floor sometimes develops a break in the floor, or sole. The division is necessary because there must be enough room for congestion and swelling before and after calving so that there will be no injury to the udder. (This is another instance of the advantage of having a practical knowledge of the functional aspects in order to make good judgments in the show ring.) Finally, the rear udder should be so placed that it fits properly with the fore udder.

The fore udder is very important in the overall udder conformation. In form, the fore udder should be moderately capacious. Its floor should continue at the same level as that of the rear udder. The ideal fore udder is moderately long and has a gentle and gradual curve upward before it blends smoothly with the body.

A strong attachment will hold a high-producing udder in place for many years and will prevent the formation of abnormal tissue and edema (collection of lymph fluids) in the udder. The term "moderately long fore udder" is used because a fore udder that is intermediate in length and has a strong, smooth

---

[1]H. D. Norman and others, "Relation of First Lactation Production and Conformation to Lifetime Performance and Profitability in Jerseys," *Journal of Dairy Science,* 64 (1981), 104–113.

[2]J. E. Honnette and others, "Prediction of Herdlife and Lifetime Production from First Lactation Production and Individual Type Traits in Holstein Cows," *Journal of Dairy Science,* 63 (1980), 816–824.

attachment is likely to wear longer on a high-producing cow than an exceptionally long fore udder that breaks more readily at the point of attachment. Here again we see the importance of the practical approach in setting up the standard.

Both fore quarters should be equal in size and shape, and, together with the rear quarters, should form a well-shaped udder. The floor of the udder should have a slight division, for the same reason described for the rear quarters. This usually indicates a strong medial suspensory ligament.

The teats should be uniformly placed under the corners of the udder floor. They should not be widely spaced because this often creates difficult machine milking. They should hang straight down from the udder floor and should be of average, convenient size and shape. This is important because the teat and the udder size and shape are often associated with good milking qualities. Pointed teats milk out only half as fast but have the lowest cell count and are less apt to contract disease. A round teat-end is the favored shape because it had the second lowest cell count and a good milking rate in experiments at Washington State University. The suspensory ligament is very important in the overall alignment.

The veining of the udder and body is no longer emphasized as it was some years ago. As a rule, prominent udder veining is an indication of high quality, but in exceptional cases distended veins may indicate lack of quality.

Udder quality refers to the condition of the secreting tissue in the udder. Good tissue is soft, pliable, and spongy to the touch, with no firm, fatty tissue or edema in the udder. Edema causes abnormal congestion at calving time and persists after cows have been fresh for a long time. A "meaty" udder is a relatively poor producer.

To a large extent, udder quality cannot be accurately estimated by observation, and even after much experience it is advisable, when judging, to handle and examine the udder. When this is not possible, udder quality and strength of fore and rear attachments can be observed best when the cows are walking. True udder quality can only be determined with precision when the udder is milked out.

UDDER ATTACHMENTS

To evaluate udder attachments, the judge must know how the udder is suspended from the body. The seven supporting structures are best demonstrated by pictures (Figs. 190 through 196) based on studies made by the research staff of the U.S. Department of Agriculture Bureau of Dairy Industry.[3] These researchers list the following means of udder suspension:

[3]W. W. Swett and others, "Arrangement of the Tissues by Which the Cow's Udder Is Suspended," *Journal of Agricultural Research,* 65 (1942), 19–43.

Figure 190 The skin tissue serves in a minor capacity to suspend and stabilize the udder. (Courtesy Dairy Husbandry Research Branch, Agricultural Research Service, USDA)

Figure 191 A sheet of fine connective tissue just beneath and loosely attaching the udder to the skin is clearly shown between points *a* and *a*. The fine subcutaneous tissue (superficial fascia) serves as an attachment between the skin and the underlying tissues. (Courtesy Dairy Husbandry Research Branch, Agricultural Research Service, USDA)

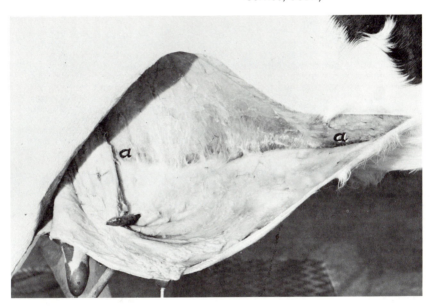

1. The skin.
2. A sheet of fine connective tissue just beneath the skin.
3. Cordlike tissue connecting the udder and body wall.
4. Fibrous and elastic tissue which forms a sling downward and forward over the udder.
5. Deep side tissue, suspended from the pelvis (bony structure of the rump) by the subpelvic tendon, envelops the udder, and is fastened to the lower side. This forms the lateral suspensory ligaments.

6. Subpelvic tendon which extends from the center of the pelvis to the center line of the udder so tissues 4 and 5 can attach to it.

7. The middle yellow elastic tissue which attaches to the body wall on one end and extends down between the two halves of the udder to divide it lengthwise. This forms the medial suspensory ligament.

The importance of the suspensory attachments for the udder is indicated by the weight they have to carry. Research workers in the Bureau of Dairy Industry at Beltsville reported an average empty weight of 52 lb for 50 udders removed from lactating cows. The udders of Holstein cows, during the early part of the lactation, averaged 73 lb. To this must be added the weight of the milk and blood in the udder. It is estimated that high-producing udders, together

Figure 192   The cordlike tissue connecting the udder and the body wall is shown at the tip end of arrow *a*. This tissue forms a loose bond between the upper surface of the front quarters and the abdominal wall. It may give way because of excessive udder weight, injury, or inherited weakness and permit a separation between the udder and the abdominal wall. This familiar "breaking away" of the front quarters is sometimes extensive enough to permit one's hand to be inserted between the cow's abdomen and the top front quarters of the udder. The condition is not too serious until the break becomes severe; udders often function normally for years after tissue No. 3 has allowed the front quarters to become loosened. The tips of arrows *b* show the very important sheet of suspensory tissue (No. 5) which envelops the entire udder. (Courtesy Dairy Husbandry Research Branch, Agricultural Research Service, USDA)

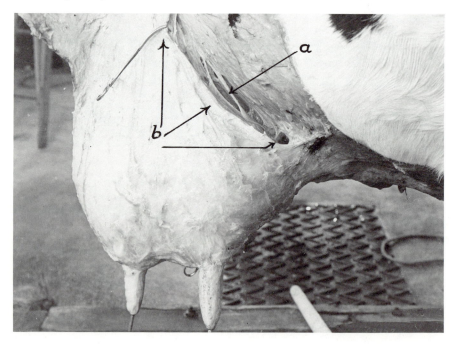

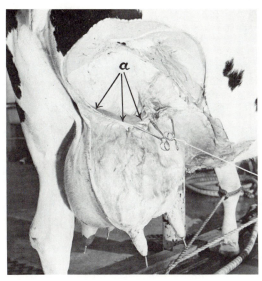

Figure 193 Fibrous and elastic tissue No. 4, known as the *lateral suspensory ligament,* forms a partial sling downward and forward over the udder. This is an important means of suspension near the surface, and it attaches the udder to the body wall and also to the inner surface of the thigh. The severed edge of tissue No. 4 is indicated by the line following the tips from the arrow arising at *a.* It is held in place after dissection by the forceps attached to the tissue. (Courtesy Dairy Husbandry Research Branch, Agricultural Research Service, USDA)

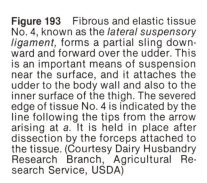

Figure 194 Deep side tissue No. 5, a tough, fibrous, suspensory tissue that almost envelops the udder. It is connected to the floor of the pelvic bones by tissue No. 6 and is fastened on the other end to the lower surface of the udder by numerous platelike attachments which pass into the gland and become part of the framework of the udder. The tissue has been severed at *a* and folded back along the line *b–b* to show how it envelops the udder. (Courtesy Dairy Husbandry Research Branch, Agricultural Service, USDA)

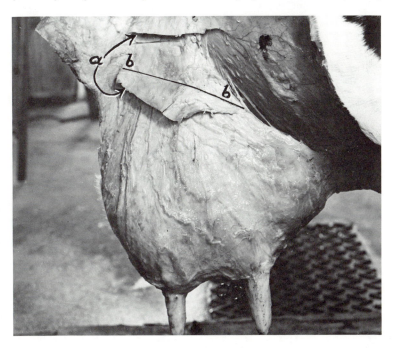

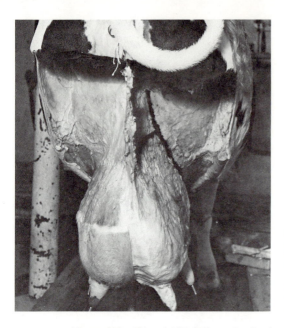

Figure 195 The subpelvic tendon resembles a cable or rope and extends from the floor of the pelvic arch and the thurls to the center line of the udder so that tissues Nos. 4 and 5 can attach to it. (Courtesy Dairy Husbandry Research Branch, Agricultural Research Service, USDA)

Figure 196 The middle yellow elastic tissue No. 7 consists of two strong sheets attached to the body wall just above the center of the udder. It extends down between the halves of the udder to form an important means of support, and it divides the udder lengthwise into distinct halves. The great strength and the almost perfect location of this middle support are shown by *a*. This is the important medial suspensory ligament. (Courtesy Dairy Husbandry Research Branch, Agricultural Research Service, USDA)

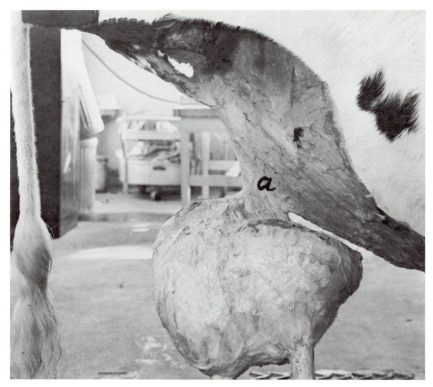

with their contents, weigh between 100 and 250 lbs. This load must be well suspended if it is to remain in proper position.

Some of these attachments may be faulty and others good, which accounts for the many different degrees of faulty udder attachments. It also explains why a partially weakened fore udder may remain in the same state for many years, or, on the other hand, may become completely broken within a year.

DEVIATIONS FROM THE IDEAL

Expected deviations from the ideal and possible defects in the mammary system are presented on the following pages. The seriousness of the defect, as evaluated by successful and knowledgeable judges, is indicated by the degree of discrimination recommended. To a large extent, the discrimination is based on the functional and economic importance of the impairment in terms of lifetime production. The final discrimination depends entirely on the degree of deviation from the ideal. By applying good judgment to the variation, and considering how much it may decrease production, it is possible to arrive at a correct placing and one that will be fairly uniform for a group of several judges.

A detailed description of these deviations is not necessary if one keeps the ideal clearly in mind. For degree of discrimination, *slight* is equivalent to 3 or 4 places; *moderate,* 8 to 10; and *serious,* 15 to 18 places in a very close class of 20 or more individuals of good type. Disqualification means that all animals in the class not disqualified will be ranked above the disqualified individual. If there are several disqualified animals, these will be ranked at the bottom of the class. The range of discrimination on faults allows for different degrees of deviation.

Deviation and Defects	*Degree of Discrimination*
1. Fore or rear attachment fairly good but not firm	Slight
2. Loose fore or rear attachment	Moderate
3. Weak udder attachment	Slight to serious
4. Broken fore or rear attachment	Serious
5. Pendulous udder (broken fore and rear attachment); broken suspensory ligament	Very serious to disqualification
6. Tilted udder with deficient fore attachment	Moderate to serious
7. Tilted udder with low-hanging rear udder and loose rear udder attachment	Serious
8. Short fore udder	Slight to moderate
9. Small udder lacking in capacity	Moderate to serious
10. Round, tight udder	Serious
11. Udder hanging or pitched too far forward	Moderate to serious

12. Too much udder showing back of rear leg   Slight to moderate
    or bulgy rear udder
13. Too little udder showing back of rear leg   Slight to moderate
14. Broken floor of udder   Serious
15. Bulging fore udder   Slight to serious
16. Light unbalanced quarters   Slight to serious
17. Quartered and halved udder   Slight to moderate
18. Teats pointing forward   Slight to moderate
19. Teats pointing outward   Moderate to serious
20. Rear teats too close   Very slight
21. Teats too close from side view   Slight to moderate
22. Widely spaced front teats   Moderate to serious
23. Teats too far on outside or too far back   Moderate to serious
24. Lack of udder and/or body veining   Very slight
25. Hard spots in udder   Slight to serious
26. Abnormal milk (bloody, clotted, watery)   Slight to serious
27. Side leak   Slight
28. Obstruction in teat   Slight to serious
29. Edema or permanent congestion in udder   Serious
    long after calving
30. Blind quarter   Disqualification

If an inexperienced judge remembers the degree of discrimination assigned to each deviation from the ideal and evaluates the particular characteristic accurately, he or she can arrive at a sound placing.

UDDER TERMINOLOGY

A judge should develop a good vocabulary for describing type characteristics and deviations observed. To accomplish this, udder terminology will be developed similar to the procedure used in previous chapters for the other points of conformation.

Similar words or sentences are occasionally included to help build up a vocabulary consisting of various expressions for giving reasons. Expressions describing favorable and unfavorable type characteristics are listed below:

Favorable:
1. High, wide, and strong (firm) (smooth) rear udder attachment.
2. Long, level, smooth, firm (strong) fore udder attachment.
3. Capacious udder extending well forward.

   4. Large udder with an advantage in capacity.

   5. Excellent size and shape of udder.

   6. Proper amount of udder showing in back of rear leg.

   7. Udder having excellent width as indicated from the rear view.

   8. Udder that is uniformly wide, from attachments to udder floor.

   9. Udder that closely approaches perfection.

  10. Mature udder, excellent in every respect, with little evidence of wear.

  11. Proper depth of udder.

  12. Supported by a strong medial suspensory ligament.

  13. More capacious udder.

  14. Evenly balanced and strongly attached udder (evenly balanced quarters) (exceptional balance of udder or quarters).

  15. Well-balanced, symmetrically shaped udder.

  16. Udder attached neatly in its junction with the body wall.

  17. Fore udder full and well developed.

  18. Longer fore udder.

  19. Udder is pliable, soft, and spongy and therefore excellent in texture and quality (high quality and no meatiness).

  20. Nearly ideal spacing of teats (uniformly or evenly spaced teats).

  21. Symmetrically placed teats of uniform and convenient shape and size.

  22. Ideally spaced teats at each corner of the udder floor (teats set neatly on the udder floor).

  23. Teats that hang perpendicularly from the quarters (to the ground).

  24. Prominent (tortuous) (distinct) (extensive) veining on udder and/or on body (barrel). (Long, large prominent body veining.)

  25. Abundant (numerous) well-defined udder veins.

Unfavorable:

   1. Weak attachment of fore (rear) udder.

   2. Broken attachment of fore (rear) udder. (Fore, or rear, quarters broken away at the attachment.)

   3. Pendulous (swinging) udder.

   4. Pear- (cup-) (funnel-) (pumpkin-) shaped udder.

   5. Low-hanging, weakly attached udder.

   6. Small udder with deficient, undersized quarters. (Small rear or fore quarters lacking fullness and development.)

   7. Udder lacking width, depth, and capacity. (Small udder with lack of capacity.)

   8. Shallow fore and rear udder.

9. Round, tight udder with limited capacity.
10. Udder cut up (divided) too much between the quarters. (Udder badly quartered and/or halved.)
11. Low-hanging rear udder. Rear (fore) quarters hanging lower than (far below) fore (rear) quarters.
12. Narrow, weak (loose) rear attachment.
13. Bulging rear udder, with too much showing behind the rear leg.
14. Deficiency of (cut under) rear udder; not enough udder behind the rear leg.
15. Steep fore udder.
16. Cuts up abruptly in the fore udder.
17. A bulging (side, floor, both) fore udder.
18. Tilted, deficient fore udder.
19. Unbalanced udder with large rear quarters.
20. Floor of udder below the hock.
21. Broken floor (sole) of the udder. Weak (broken) suspensory ligament.
22. Quartered, halved, cleft udder. (Severe cleft in floor of the udder.)
23. Unbalanced (light) (heavy) (enlarged) (hard) (slack) (weak) (deficient) quarter.
24. Udder seriously lacking in quality. Lumpy (meaty) (hard) udder.
25. Edema or congestion of udder long after calving. (Udder with permanent congestion.)
26. Irregular (small) (funnel) (pencil) (long) (tapering) (defective) (large) (bottle-shaped) teats.
27. Teats hanging too close together when seen from the side.
28. Teats set too far on the outside of the quarter (too far back on rear quarters).
29. Teats too wide in front.
30. Teats pointed (strut) forward (outward).
31. Lack of udder veining.
32. Side leak or blemished teat.
33. A blind, nonfunctional quarter.

CONSIDERATION OF AGE AND STAGE OF LACTATION

Udder shape and size change with age and stage of lactation. These factors should always be taken into consideration for proper evaluation of the udder.

We expect younger cows to have less capacious but more firmly attached udders than older cows (Figs. 206, 207, 371, and 372). Hence, an udder that is somewhat small and shallow but very firm in fore, rear, and median support of udder is highly desirable in a 2-year-old. An udder of similar capacity in a mature cow would be penalized for lacking udder capacity. A large, deep udder having attachments that are a bit loose would be acceptable on an older cow but would be severely penalized on a 2-year-old. Evaluation of wearing qualities and strength of udder attachment should be much more critical in younger cows than in older animals. If a young cow shows too much wear in her udder, her future is not promising, whereas some wear can be expected in older cows, especially those having exceptional dairy qualities and outstanding production.

Stage of lactation has a definite effect on the capacity, shape, quality, and apparent firmness of udder attachments. A good judge can make a reasonably accurate estimate if the exact information is not available.

During the early months of lactation, productive cows should have large, capacious udders. They are often referred to as having "bloom of udder." Any

A

Figure 197 Udders are assigned so much importance on the score card and in a breeding program because they represent the working end of the cow. (A) Jersey cows that are all daughters of a sire who had a high predicted difference in production and type. This bull made a real contribution to breed improvement. (Courtesy Sonshine Jersey Farm, Brashear, TX) (B) Daughters of a Holstein bull that transmitted high production and superior udders, legs, feet, and overall conformation.

B

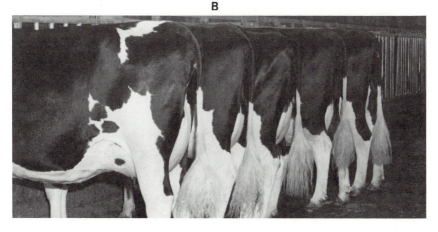

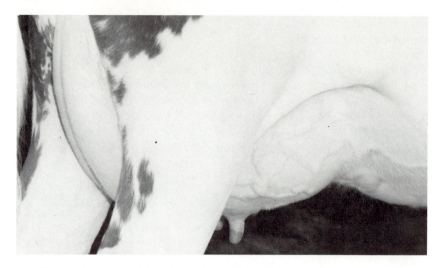

Figure 198 An Ayrshire udder, winner of the All-Breed Best-Udder Class at the Royal Winter Fair in Canada. (Courtesy Ayrshire Breeders Association, Brandon, VT)

tendency toward weakness of attachment will be evident during this period. During the later stages of lactation, when the cow is producing less milk, she generally shows less capacity but more quality and smoother appearing udder attachments. Lack of udder capacity should be penalized more in fresher cows than in stale cows; lack of firmness of udder attachments, deep udders, or lack of udder quality should be penalized more in stale cows. The experienced judge takes these factors into consideration when evaluating cows in various stages of lactation. Judging a dry udder requires special techniques and observations and is discussed in Chapter 16.

Figures 198 through 202 show outstanding udders in each of five dairy breeds. Note the common denominators for well-attached fore udder, high and wide rear udder, firm support by the suspensory ligament, and teat size and arrangement.

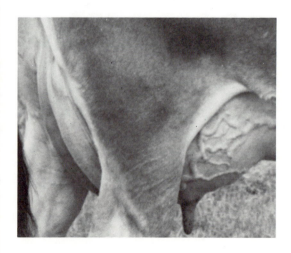

Figure 199 Outstanding Brown Swiss udder which was the winner of the best udder in class at the National Show. (Courtesy Howard Voegeli, Monticello, WI)

Figure 200 This almost perfect udder on a Guernsey cow was many times the winner of the best-udder award at state and national shows. Note the firmness and smoothness of attachments and the good suspensory ligament which carries the udder up tight for a marvelous overall shape. (Courtesy Woodacres Farm, Princeton, NJ)

Figure 201 Holstein udder having outstanding capacity, a firm medial suspensory ligament resulting in a level udder floor, and almost ideal teat placement. As cows with heavy production reach mature age, a large udder is normal, but the attachments (fore, rear, and medial suspensory) should be firm, as displayed by this cow pictured when she was milking 168 lb per day. She was classified E across on udder and 3E 96 overall. She was first in 305 and 365 days National Leader for fat. Her cow index was +1002M +.14% +61F +$169, CTPI +702. At 7 yr 3 mo in 365 d she produced 41,320 lb M, 4.4%, and 1817 lb F.

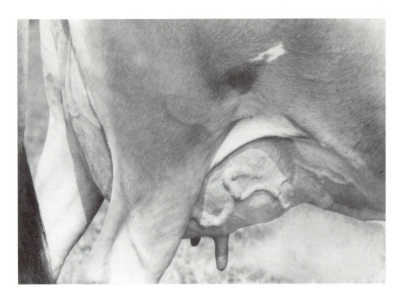

Figure 202   A superb udder on a 9-year-old which won the Best Udder for Aged Cows Class at the National Jersey All-American Show. (Courtesy American Jersey Cattle Club, Columbus, OH)

Figure 203   George Trimberger points out the characteristics that make this an almost ideal fore udder on this 3X All-American, Ex 96 Holstein cow with over 20,000 lb of milk and a high lifetime performance. (Courtesy Gray View Farm, Union Grove, WI)

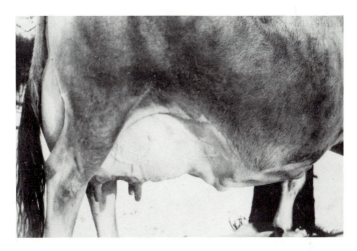

Figure 204 Udder of Jane of Vernon as a young cow (A) and at 11 years of age (B). This cow had one of the best-shaped and best-quality udders ever known to any breed. It was superior to the ideal Brown Swiss udder and can serve as the ideal for any breed. A Grand Champion at the National Dairy Cattle Congress for 5 years in succession, her production included a world's record as a 4-year-old, with 1076 lb of fat. She had two records of over 1000 lb of fat. Through her transmitting ability she had a powerful influence on udder and body type improvement of the Brown Swiss breed. See Figs. 2 through 5 for this cow at 3, 4, 11, and 15 years of age. (Courtesy Brown Swiss Breeders Association, Beloit, WI)

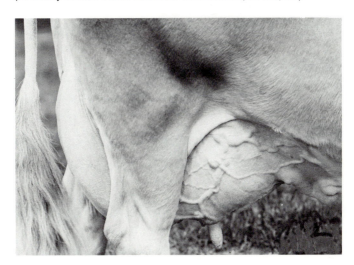

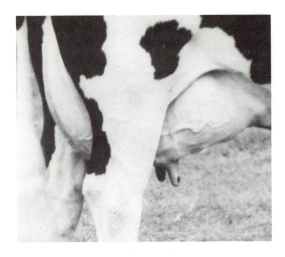

Figure 205   The right kind of udder on this Holstein cow. She carried this almost ideal udder to a very advanced age. She classified 96 and was selected as the All-Time All-American 4-year-old. She was nominated for All-American for 6 consecutive years. (Courtesy Hanover Hill, Armenia, NY)

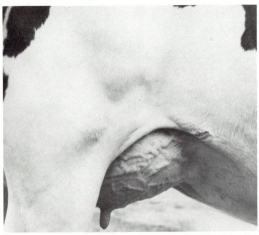

Figure 206   The right kind of udder on a 2-year-old. For the first lactation, the udder should not be too large or too deep. This comes with time, and due allowance should be made for the expected high production. (Courtesy Collins-Crest Farm, Perry, NY)

Figure 207   This superb mature udder is on the All-Time and now Reserve All-Time All-American Aged Cow, Ex 97, 3X All-American Aged Cow with three records over 25,000 lb of milk and three records over 1000 lb of fat. (Courtesy Paclamar Farm, Louisville, CO)

Figure 208 A faulty udder that is loose in both fore and rear attachment and that has a weak medial suspensory ligament, permitting too much depth of udder (a serious discrimination).

Figure 209 The rear udder is pitched forward too far, and the fore udder is broken, but the medial suspensory ligament prevents the udder from going down too far (a serious discrimination).

Figure 210 This udder is pitched forward too far, and the fore udder is slightly bulgy, but the attachments with the suspensory ligament are still serving a useful purpose (a moderate discrimination).

Figure 211 A tilted udder with a poorly attached rear udder, a weak suspensory ligament, and a deficient fore udder (a serious discrimination).

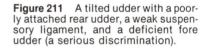

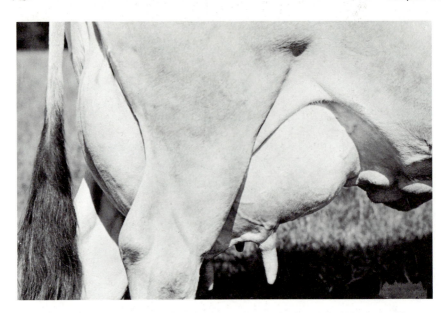

Figure 212   Good udder of Brown Swiss cow produced 31,283 lb of milk, 1379 lb of fat for a national record. (Courtesy Lee's Hill Farm, New Vernon, NJ)

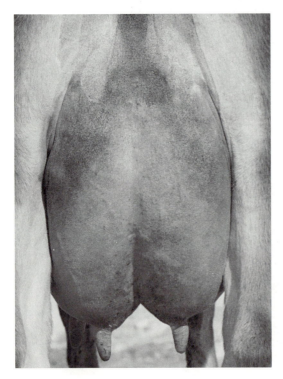

Figure 213   The division or cleft showing on this E-91 Jersey cow with
21,050 lb M and 1043 lb F in 305 d on
2X at 3 yr, 3 mo indicates an outstanding medial suspensory ligament.
This udder support holds the udder up
tight against the body and divides the
udder lengthwise. Identification,
evaluation, and selection for a superior
medial suspensory ligament has
resulted in much progress for improvement of functional udders that wear.
(Courtesy Jack and Judy Earnest,
Fredericktown, OH)

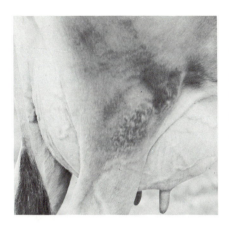
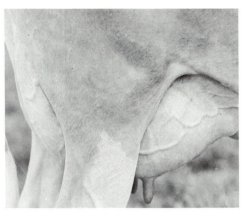

Figure 214   Excellent udders on two Jersey sisters with outstanding production. The udder on the left achieved a new national record of 17,496 lb of milk and 829 lb of fat in 365 days as a senior 2-year-old. This cow averaged 47.9 lb of milk per day during the year for her first record, with the high day 10 lb above this. The udder was photographed at the completion of her record, after she had produced 21 times her weight in milk during the first 12 months. Without a dry period, she made a high record as a senior 3-year-old and followed this with a new national record for both milk and fat as a senior 4-year-old. The sister's production has been almost as good. The females in this pedigree include a number of great lifetime milkers, and the 14 nearest relatives are classified excellent. This cow is in the fifth generation of excellent cows. (Courtesy Marlu Farms, Lincroft, NJ)

Figure 215   A dam, on the left, and her daughter, with udders that classified excellent and broke the national records for production. The udder of the dam set a new national record for junior 2- and 3-year-olds (18,493 lb of milk, 4.4%, 811 lb of fat on 2X). The daughter with an improved udder set a new national record for senior 3-year-olds (15,488 lb of milk, 5.1%, 779 lb of fat actual on 2X). This combination of type and production in successive generations was the result of a carefully planned breeding program. (Courtesy Vista Grande Farm, Cropseyville, NY)

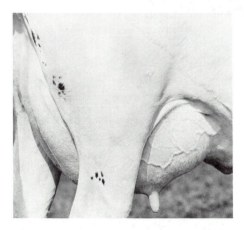
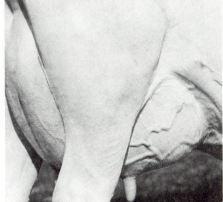

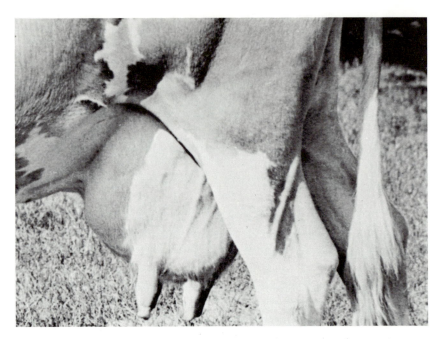

Figure 216 High-producing cow showing excessive wear from high production. Note her broken fore udder and rear udder, which will greatly lessen her ability to continue the high production as she gets older (a serious discrimination).

Figure 217 Excellent udder veining on a high-producing Holstein cow. This is a continuation of prominent and tortuous body veining.

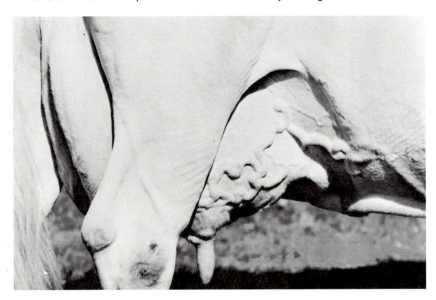

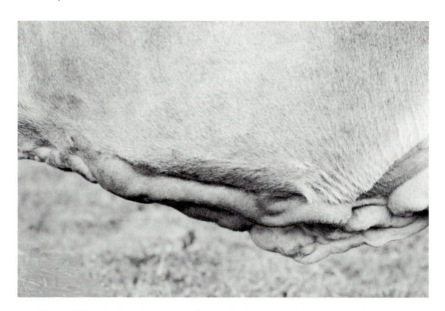

Figure 218 Body veining on a Brown Swiss cow with unusually high production records.

Figure 219 An outstanding, firmly attached, properly shaped udder having a good medial suspensory ligament. (Courtesy Ayrshire Breeders Association, Brandon, VT, and Pinehurst Farms, Sheboygan Falls, WI)

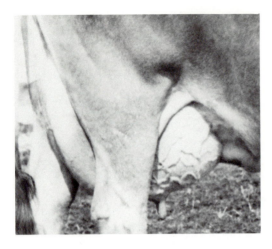

Figure 220 Superior udder of a Brown Swiss cow that milked 27,149 lb of milk at 11 yr, 2 mo and produced over 20,000 lb of milk six times, over 1000 lb of fat three times, and over 200,000 lb of milk lifetime. This udder is on a 4E cow that was nominated All-American in every milking class, including a nomination at 12 years of age. She was the national performance winner. (Courtesy Leon Button, Rushville, NY; Hanover Hill, Avon, NY; Brigeen Farm, Turner, ME)

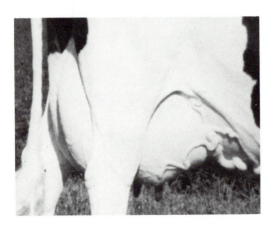

Figure 221 This almost perfect udder is on a 3E 97 high-producing Holstein cow. Note the levelness, preferred length, strength, and smoothness of fore attachment combined with a strong medial suspensory ligament holding the udder in the right place (photo reversed). Courtesy Bob and Kay Miller and family, Mil-R-Mor, Dundee, IL)

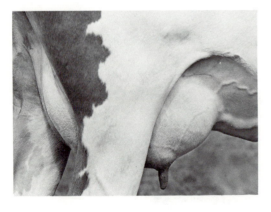

Figure 222 A fore udder that cuts up too abruptly in front of the fore teat. It also lacks strength and smoothness of attachment (a moderate discrimination).

Figure 223 Short, loosely attached fore udder that is too rounded on each side of the teat with strutting front teats (a moderate discrimination).

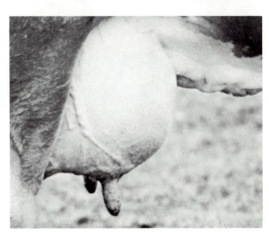

Figure 224 Rounded udder with a broken fore attachment and a loose rear attachment. The gaping fore udder should receive a serious discrimination.

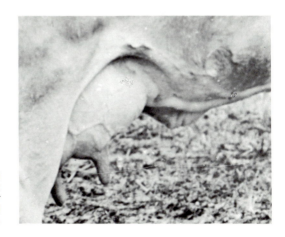

Figure 225 Tilted fore udder and loose rear udder attachment that pushes the rear udder ahead too far and causes the teats to point forward (a moderate to serious discrimination).

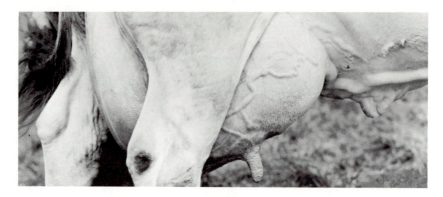

Figure 226 Poorly attached, tilted udder that lacks in quality with teats too far on the outside of the fore udder (a serious discrimination).

Figure 227 Superior udder of a Brown Swiss cow. Note the wide, strong, smooth attachment of the rear udder. This cow was several times a winner of the best-udder class at national and international shows. (Courtesy Lee's Hill Farm, New Vernon, NJ)

Figure 228 The exceptionally wide spacing of teats on this udder is admirable, but the rear udder bulges and extends out so far that it has to cut back at the attachment. Also, the fore udder is longer than necessary. An udder that is as long in attachment as this fore udder is susceptible to severe "breaking away" (a moderate discrimination).

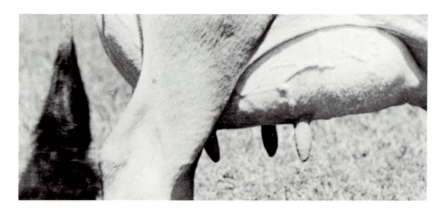

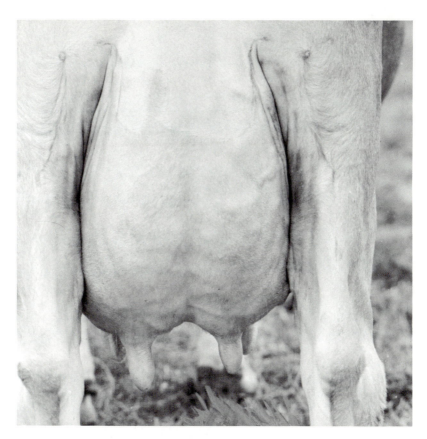

Figure 229 Rear view of the udder shown in Figure 227. This rear udder, with its wide, smooth, strong, and even attachment and its perfect size and proportion can serve as an ideal. (Courtesy Lee's Hill Farm, New Vernon, NJ)

Figure 230 Narrow, pinched, loosely attached rear udder of a young cow. Udders of this sort do not hold up and receive serious discrimination.

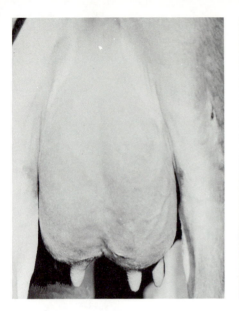

Figure 231 Slight unbalance can be observed between the rear quarters. Since no udder is perfectly balanced between the quarters, there is no cut or discrimination for this slight deviation.

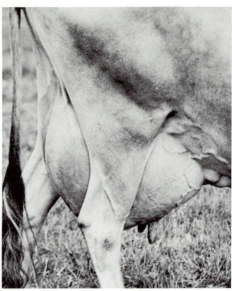

Figure 232 Broken rear attachment permits the rear udder to drop, causing an uneven, tilted udder floor (a serious discrimination).

Figure 234 Rear udder with a good attachment, but the fullness (straight floor of udder from rear view) between the teats is objectionable because there is no room for the inflammation usually present in an udder before and after calving. Often a filled sole of the udder accompanied by excessive depth of udder indicates a weakening of the medial suspensory ligament. This udder is flat on the floor though not deep (a slight to moderate discrimination).

Figure 233 No discrimination is made for small extra teats if they appear nonfunctional. Good management procedures, however, direct that extra teats be removed before 1 year of age.

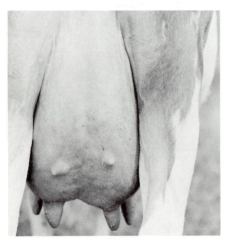

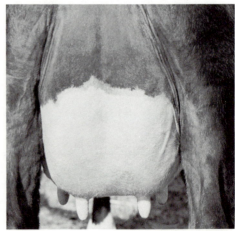

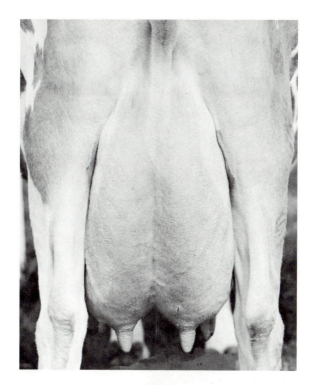

Figure 235 Division between these rear quarters is fairly pronounced and indicates a strong medial suspensory ligament. It is considered ideal. This condition is preferable to that illustrated in Fig. 234, and the udder is sound from a functional standpoint. Therefore, it should be considered an advantage and given proper credit.

Figure 236 A good, practical working udder. This Brown Swiss cow was three times National Total Performance winner and has a lifetime production of over 194,000 lb of milk and 7900 lb of fat. (Courtesy St. John's Dairy, Glendale, AZ)

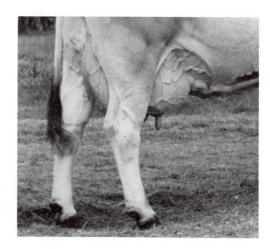

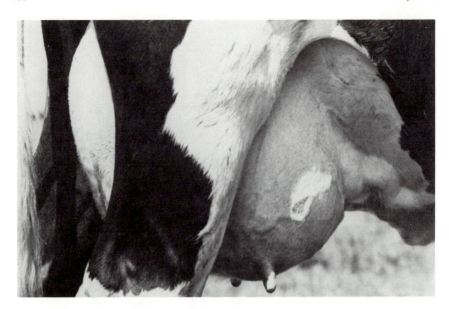

Figure 237   An udder with all attachments beginning to break. The broken medial suspensory ligament permits the floor of the udder to drop and push the teats to the side of the udder (a very serious discrimination).

Figure 238   This cow has a deep udder with a broken medial suspensory ligament that allows the udder to drop well below the point of the hock (disqualification).

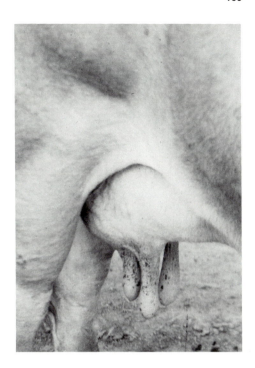

Figure 239 A small, ill-shaped udder with extremely large bulbous, undesirable teats (a very serious discrimination).

Figure 240 A pendulous udder with a broken medial suspensory ligament plus a broken fore and rear attachment (a disqualification).

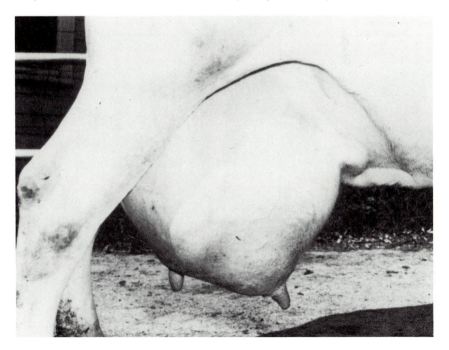

Figure 241   The broken medial suspensory ligament, together with the other broken attachments, drops the floor of this edematous udder and pushes the teats to the side of the udder (a disqualification).

Figure 242   Side and front views of a broken edematous udder that is useless for milk production. Note the broken floor caused by a broken medial suspensory ligament (a disqualification).

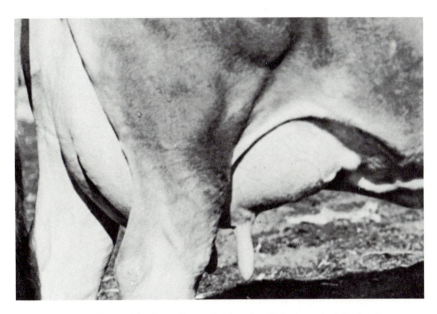

Figure 243 This udder is well attached and satisfactory, but the teats are too long (a moderate discrimination).

Figure 244 These teats are too pointed and tapered, which causes difficulty during milking (a moderate discrimination).

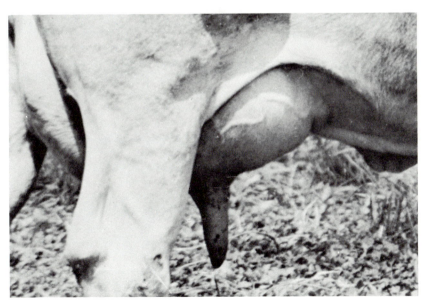

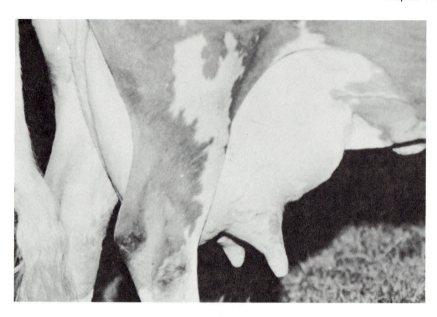

Figure 245   Funnel-shaped teats (moderate discrimination) on a tilted, poorly attached udder (a serious discrimination).

Figure 246   Funnel-shaped teats placed too close together on a tilted udder with poor attachments (a serious discrimination).

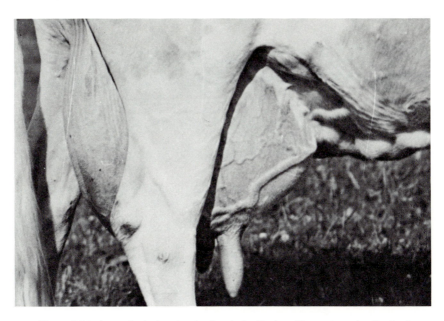

Figure 247 Large teats (moderate discrimination) and large, loosely attached udder (serious discrimination) make a very objectionable combination.

Figure 248 Loosely attached udder with teats close together and pointing forward (a serious discrimination).

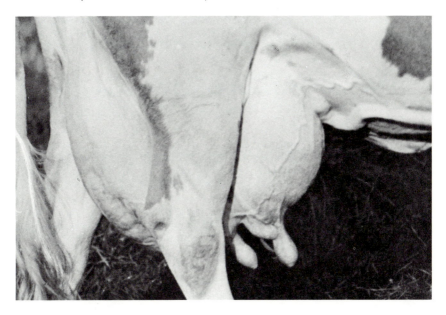

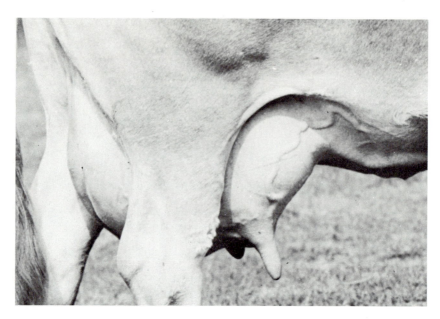

Figure 249 Small, poorly attached, quartered udder. The short fore udder has teats that point forward and outward (a serious discrimination).

Figure 250 Broken fore and rear udder attachments. This resulted in an ill-shaped, swinging udder that is very susceptible to injury (a serious discrimination).

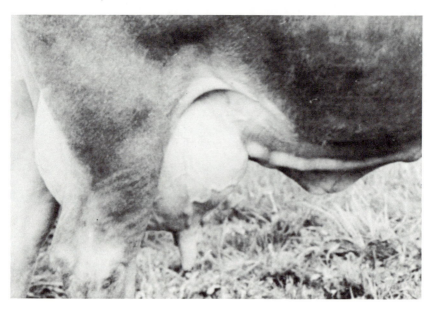

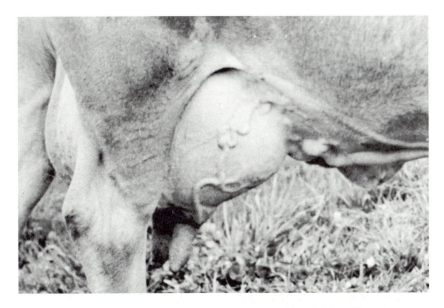

Figure 251 Broken fore and rear attachments permitted this udder to drop in front so that the fore teats point backward. This udder is readily susceptible to injury and must be severely criticized (a very serious discrimination).

Figure 252 Broken fore and rear attachments plus a broken medial suspensory ligament make this udder pendulous (a disqualification).

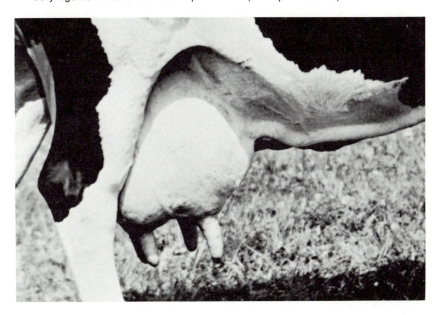

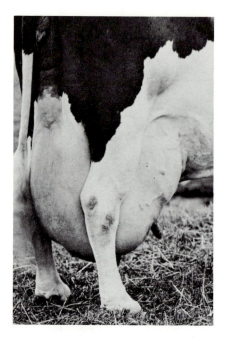

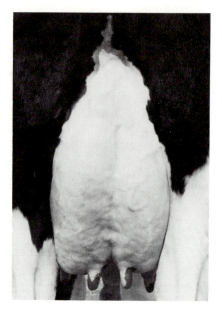

Figure 253 Severe break in rear udder, combined with permanent congestion, produced this ill-shaped udder, which should be penalized to the extent of disqualification.

Figure 254 The crease in this rear udder is very desirable because it indicates a highly developed center suspensory ligament. Note the height, width, and strength of rear attachment. (Courtesy King Syndicate, Millard Brink, Ithaca, NY)

Figure 255 (A and B) Attractive udder on a high-producing Holstein 5 Ex 97, Reserve All-American first four lactations and then twice All-American Aged Cow. The side and back views indicate the soundness of this udder after a production of 32,471 lb of milk, 3.7%, and 1215 lb of fat at 5 yr, 7 mo, breaking the Wisconsin state record for milk in 305 and 365 days. (Courtesy Crescent Beauty Farms, Fort Atkinson, WI)

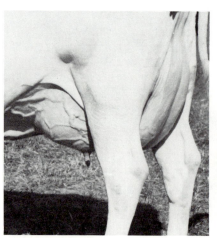

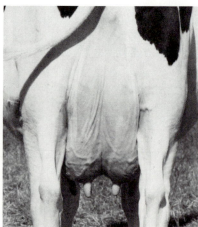

12
DAIRY CHARACTER

In general, cows that are lacking in dairy character are correspondingly poor producers, and cows of outstanding dairy character make high production records. There are exceptions to this, but in judging it is best to play the odds in favor of making a correct evaluation and decision. Numerous researchers have reported that there is a highly positive correlation (from 0.45 to 0.60) between dairy character and milk yield. This close association between production and dairy character indicates that dairy character should receive considerable emphasis in dairy cattle judging. A judgment based on correct analysis of dairy character rarely fails the judge in the task of selecting a useful, high-producing cow.

Heavy, coarse cows, seriously lacking in dairy character, usually convert a proportionately larger amount of feed into body maintenance and fatty deposits than do cows of exceptional dairy quality. From this, it is apparent that a poor producer on a good ration will accumulate fat and fleshiness, and that as her body takes on weight she will require more feed for maintenance. These fleshy animals usually produce considerably less milk than do sharp, angular cows of the same skeletal size but weighing several hundred pounds less.

Dairy quality does not imply a thin, emaciated condition, for dairy character is at its best when combined with strength, which provides staying power for sustained high production year after year. A frail cow cannot meet the energy requirements of cumulative high production. A certain amount of fleshing and strength of frame is necessary to provide outstanding wearing

167

qualities. One should think of dairy quality as a combination of refinement and strength.

It is now generally recognized that fat deposits and coarseness interfere with the functioning of the vital organs and the secreting tissue in the udder of a dairy cow. In contrast, a spare, lean frame, indicative of a highly developed dairy character, is usually a sign of heavy milk production. The authors have made a special study of the form and dairy character of high lifetime producers and of cows in the 100,000-lb production class at shows. Almost without exception, such cows are angular, with a very lean neck and sharp open conformation, showing a good frame with ample dairy refinement.

IDEAL DAIRY QUALITY

Dairy character is best indicated by a clean-cut, angular, open conformation and a strong, refined appearance, with freedom from coarseness and excess flesh throughout. The head should be clean-cut and possess plenty of character; the neck should be long and thin and should blend smoothly; the withers should be sharp, angular, and well defined; the vertebrae of the spine should be open, prominent, and clean-cut; the rib bones should be flat, wide, long, and far apart to give openness of body; the hooks or hips should be very prominent, sharp, and well defined; the pins should be sharp; and the thighs should be thin and incurving.

Normally, a dairy cow that has these characteristics turns just about all of the feed consumed over and above maintenance into milk production rather than body fat. Condition, which designates the amount of fat and fleshing that an animal carries, is important in evaluating dairy character. Cows in late lactation should not be severely penalized for not being quite as sharp and angular as cows in peak lactation, for they generally regain some or all of the weight lost during peak lactation when they were in negative energy balance (using more energy than they were consuming because of high milk production and therefore losing weight). Coarseness should be penalized in stale as well as fresher cows. A long, clean neck is usually associated with a high degree of dairy character, even in a cow that is a bit fleshy due to advanced stage of lactation. Heifers nearing freshening should not be expected to be as sharp and clean as younger calves. The experienced judge knows how to make due allowance for the influence of late lactation, the dry period, and heifers in late gestation.

SOME USEFUL DESCRIPTIVE TERMS

1. Clean-cut head.
2. Coarse (thick) (short) head.
3. Long and clean neck.

4. Clean throat.
5. Short neck.
6. Coarse (thick) neck.
7. Coarse shoulders.
8. Clean through the front end.
9. Clean brisket.
10. Coarse brisket.
11. Heavy dewlap.
12. Sharp (clean) withers, topline, rump.
13. Prominent vertebrae, hips, and pins.
14. Fleshy rump.
15. Coarse hips and pins.
16. Patchy pins.
17. Clean (flat) bone.
18. Coarse (rounded) bone.
19. Clean (incurving) thighs.
20. Thick (rounded) thighs.
21. Open ribbed.
22. Tight ribbed.
23. Exhibits sharpness and cleanness throughout her body.
24. Is angular throughout her body.
25. Is thick and coarse throughout her body.
26. Has a clean, milky appearance.
27. Is free of excess flesh.

POSSIBLE DEVIATIONS FROM THE PREFERRED TYPE

Deviations from the ideal dairy character include, in varying degrees, the following: short, heavy, and coarse head; short and thick neck; heavy, coarse, and thick shoulders with broad, heavily covered withers; closely spaced, poorly defined, and heavily covered vertebrae; rounded hips that carry excess fleshing; pins lacking in sharpness and covered with meat and fat, sometimes in patches; thighs that are thick, fleshy, and straight rather than incurving. In general, an animal that is seriously lacking in dairy character appears coarse, heavy, patchy, overconditioned, and far too compact in body conformation.

Figures 256 through 278, which illustrate dairy character or the lack of it, should help the prospective judge learn how to evaluate dairy character and attain proficiency in judging.

INCREASED EMPHASIS ON DAIRY QUALITY

More emphasis on dairy quality was justified because of the positive correlation between dairy quality and high production. This was highly acceptable to dairy cattle breeders because it involved a practical approach. At the same time there was severe discrimination against fat or over conditioned dairy cattle especially in young animals. Experimental results showed detrimental effects on production, longevity, lifetime production, a change in udder type, (from fatty deposits), and efficiency of production, when young dairy animals were extensively over conditioned. In older individuals the harmful effects from extra fat, especially for dry cows, was reflected in the increase of the downer cow syndrome and other metabolic disturbances during the early part of lactation. General coarseness with coarse bone structure, coarse front end and lack of femininity, is a severe discrimination. It reflects lower yields and a less useful cow.

Figure 256   This EX 93, twice All-American Guernsey cow looks like a producer. Note her clean-cutness, sharpness, and open conformation throughout. Her deep, open rib and body veining give her a high rating in dairy character. She lives up to this evaluation as follows:

1y–11m	2X	365 d	17,610 lb milk	4.4%	781 lb fat
2y–11m	3X	365 d	26,090 lb milk	6.7%	1737 lb fat

World's record for fat.
She was the Total Performance winner at all three national shows the same year and was selected the Grand Champion at two of these. Her MC Deviation is + 8,348 lb M and + 473 lb F. Her CIT is + 2.6 and CPI + 146. (Courtesy Elm Spring Farm, Lynn, IN)

Figure 257 A sharp and angular Guernsey cow that was a high producer and much appreciated in the show ring. Classified Excellent 7X, her highest score was 94.5. She was National Grand Champion 3X and All-American Aged Cow for 5 successive years. Note her superior udder, legs, openness in body, and sharpness (quality) throughout. (Courtesy Woodacres Farm, Princeton, NJ)

Figure 258 Study the conformation of this cow to get in mind a short, thick, dumpy cow (a serious discrimination). She produced 11,000 lb of milk.

Figure 259   In contrast to the Guernsey shown in Fig. 257, this cow lacks dairy character. She carries too much flesh; has a short, blocky head; is wide and flat on top; is close in rib placement; and is heavy and thick throughout, as indicated in the neck, shoulders, loin, rump, and thighs. (Courtesy American Guernsey Cattle Club, Peterborough, NH)

Figure 260   This high-producing Holstein cow has many of the characteristics listed for dairy character, but she lacks sufficient strength to take the stress of high production. She appears fatigued and is falling apart at an age when she should be in the prime of her life. The tired look about this cow is not due to sickness but to a lack of strength and a lack of the physical characteristics of a long and useful life. The feet and legs often are the first to reflect stress.

Figure 261 A Reserve All-American Ayrshire Senior Calf that possesses outstanding size and strength and exceptional quality as indicated by her long, lean neck and sharp topline. (Courtesy Elmknowl Farm, West Kingston, RI)

Figure 262 A thick, overconditioned Brown Swiss heifer seriously lacking in dairy character. Note the coarseness of head, neck, and shoulder that is also carried throughout her body.

Figure 263 This All-American Brown Swiss Cow 3E across signifies the ultimate in dairy character. Note the sharpness throughout her body combined with strength. All of her qualities, including the tremendous body and udder veining, are indicative of her outstanding production. At 5 years, 10 months she produced 23,250 lb of milk, 4.3%, 1002 lb of fat. (Courtesy Leon Button, Rushville, NY, and Coventry Swiss, Pavilion, NY)

Figure 264 The same cow as in Fig. 263 at 12 years of age. She is 4E overall and 4E on mammary. At 11 years, 2 months she produced 27,149 lb of milk, 4.1%, 1101 lb of fat in 365 days. She advanced to 5E and at 15 yr she produced 27,362 lb milk, 4.2%, 1144 lb fat. Her lifetime was over 240,000 lb milk and 10,000 lb fat. This cow has been nominated All-American in every milking class. She was the total performance winner (production and type) at the National Show at 12 years of age. She has six records over 20,000 milk and three records over 1000 fat. This cow was an All-American at 2 years and was nominated All-American at 12 years. (Courtesy Leon Button, Rushville, NY; Hanover Hill, Avon, NY; and Brigeen Farm, Turner, ME)

Figure 265 Strong heifer, but somewhat lacking in sharpness. Her dairy quality is not outstanding. It is usually possible to predict a heifer's appearance in milking form (see Fig. 266).

Figure 266 Here the heifer pictured in Fig. 265 is shown as an All-American 3-year-old. Although of strong conformation, she is deficient in dairy character, just as she was earlier. This can be better appreciated when she is compared with the cows shown in Figs. 267, 268, and 269.

Figure 267 Outstanding dairy character in a high-producing, All-American Cow just before freshening. The additional fleshing is necessary at this stage, and a good judge can visualize how the cow will look during heavy production (see Fig. 268) (Courtesy Woodbourne Farms, Dimock, PA)

Figure 268 The same cow shown in Fig. 267, but as she appears in working condition. Her great dairy character is a true indication of her productive capacity. She has a series of good records on 2X milking (21,179 lb of milk, 3.9%, and 821 lb of fat at 6 years, 9 months of age). (Courtesy Woodbourne Farms, Dimock, PA)

Figure 269 This 3-year-old cow was so outstanding in dairy character and other points of conformation that she won first prize at the National Show and was voted All-American as a 2-, 3-, and 4-year-old. Note the sharpness; openness of body, especially about the ribs; clean-cutness; and strength throughout her body. (Courtesy Harvey A. Nelson and Sons, Union Grove, WI)

Figure 270 A four-time All-American Holstein with an almost perfect udder. Her dairy character, however, is not as pronounced as that of some of the other cows illustrated and by present-day standards, this cow is severely criticized for lack of dairy character. (Courtesy Lavacre Holstein Farm, Modesto, CA)

Figure 271 This National Grand Champion, All-American Aged Cow, classified 2E 96, was all dairy from the top of her nose to her toes. The same year that she was All-American she produced, at 7 years, 3 months on 2X, 36,490 lb of milk and 1164 lb of fat. Some breeders referred to this cow as "dripping wet." (Courtesy Paclamar Farms, Louisville, CO)

Figure 272 Note the dairy quality and strength of this 2E 95 Holstein cow. Also note how this cow is similar to the one in Fig. 271 and is a direct contrast to the two cows in Figs. 274 and 275.

Figure 273 This 3E 97 cow is good in all departments. Since she was at the head of the production list and later had the highest index, it is appropriate to show a picture of this four-time All-American and All-Time All-American Aged Holstein Cow in the chapter on Dairy Character. Note her sharpness about the front end, at the withers, hooks and pins; together with the incurving and clean thighs; an unusually good udder plus prominent body veins; with an open, deep-ribbed body and dairy character at its best. Her production records follow:

2y-9m	364d	32,803 lb milk	3.4%	1101 lb fat
4y-7m	328d	35,517 lb milk	4.0%	1411 lb fat
5y-9m	365d	44,143 lb milk	3.8%	1698 lb fat
7y-7m	365d	48,731 lb milk	4.2%	2028 lb fat

Number 1 cow CTPI at +1088 and Cow Index +2,185 lb M
+.14% +106 lb F +$318

As All-Time All-American Aged Holstein Cow she is the highest producer of all All-Americans and has the highest production among the 54,843 tested daughters of her herd-improving sire, Elevation. (Courtesy Woodbine Farms and Romandale Farms Ltd., Arville, PA and Unionville, Ontario)

Figure 274 This cow is seriously lacking in dairy character, and her rear udder hangs too far forward. Note the thickness and coarseness about the front end of the cow, especially at the point of shoulder.

Figure 275   This coarse, overconditioned cow seriously lacked in dairy character at all stages of her lactation. This and her shallow heel of the fore and rear feet are serious discriminations.

Figure 276   In contrast to the cow shown in Fig. 257, this animal takes on considerable weight during the dry period. She should be criticized for lack of dairy character and the weight of almost 1700 lb, which is a heavy load on the bones of her legs.

Figure 277   Bulls, like cows, can excel in dairy character, as demonstrated by this sire. Due allowance is made for masculine development.

Figure 278   Holstein bull of outstanding dairy character is indicated by open body conformation and sharpness throughout. He is bred for production and sound type and is transmitting these qualities to his offspring. (Courtesy Millard Brink, Ithaca, NY)

13
GENERAL APPEARANCE

Logically, general appearance should not be discussed until all the individual parts have been described. The reader would then evaluate the assembly of the various parts as a whole. A total of 35 points is allotted to general appearance on the score card for cows; a total of 55 points is allotted for bulls. Feet and legs are included under general appearance for both sexes, but they are allotted more points for the bull. In addition, the points listed under general appearance include breed characteristics, stature, front end, and back.

General appearance, in broad terms, is listed on the score card as attractive individuality, with femininity (masculinity), vigor, stretch, scale and harmonious blending of all parts with impressive style and carriage. A top score in general appearance also requires a nearly perfect skeletal structure that is built for smooth wear and is covered with strong, smooth muscles reflecting outstanding general health. When these qualities are combined with specific breed character, which is assigned 5 and 8 points on the cow and bull score card, respectively, one can get a general idea of how the various parts of an animal should fit together. The flexibility allowed for each breed under general appearance has made it possible for a unified score card to be successful and universally accepted (see score care breed characteristics in Chapter 3).

Reference is made to Chapters 6, 7, 8, and 10, which are Head and Neck Characteristics, Shoulder Conformation, Judging Feet and Legs, and Toplines, respectively. All are an integral part of General Appearance summarized in this chapter. All of these should be understood completely in order to comprehend this part of the score card and apply its importance to precision judging.

It should also be recognized that general appearance is associated with and relates to the separate parts of dairy character and udders. Although independent of each other, it is important that the specifications for general appearance are not antagonistic or opposite to those needed for dairy character, body capacity, and udder. Important here are angularity, sharpness, upstandingness, and open conformation throughout. One characteristic that has not been put in proper focus is stature; it will be discussed next.

STATURE AND SIZE

Stature and size denote different characteristics. *Stature* refers to height at withers to give the individual upstandingness. Many good commercial dairy cows are 50 to 55 inches high at the withers. Some Holstein cows are 58 to 62 inches high, with 60 to 61 inches a common height for winners at major shows and many of the high-scoring Excellent cows.

It is a distinct advantage to have the udder and other parts of the cow carried higher in upstanding cows. It results in cleaner cows and less udder injury, provides better sanitation for a superior product, and avoids mastitis. This decreases veterinary expense and avoids the accompanying decrease in milk production.

A word of caution on stature should be introduced here. Leggedness; awkwardness; narrow front end, especially chest; and deficiency in the heart region need to be avoided. A functional, easy-moving type of cow that blends smoothly into the herd needs to be selected. The cow pictured in the last figure for this chapter exemplifies this kind in blending, stature, and size. The outstanding general appearance of this cow, together with her other good points, worked in her favor when she was voted All-Time All-American 2-Year-Old, All-Time All-American 3-Year-Old, and Reserve All-Time All-American 4-Year-Old.

Size

Cows can be too big or too small to do the job in a practical herd, where the functional kind have an advantage for a smooth, efficient operation. *Size* refers to space occupied or magnitude, usually expressed in weight for cattle. This standard has been somewhat controversial for years.

An eighteen-year breeding project at the University of Minnesota provides the best reliable information on size of cow and production achievements for total production and efficiency. All forage and grains for each cow were weighed daily. Small Holstein cows ranged from 900 to 1500 pounds, and large Holstein cows weighed 1200 to 2000 pounds at the same stage early in lactation.

In the Minnesota experiment, considerable progress was made in breeding efficient, high-producing small and large cows at an actual DHI rolling average of 20,655 lb milk and 720 lb fat in the research herd. There was no significant difference in milk and fat production between large and small cows. The large cows gave slightly more milk and the small cows slightly more fat, but the production of 4 percent fat-corrected milk was equal. Large cows ate more total feed, while small individuals ate more feed per unit of body weight. This gave the small cows an advantage because they produced comparable amounts of milk from less total feed. The salvage value of the large cows was considerably higher and birth weight of calves was significantly greater.

CONCLUSIONS ON STATURE AND SIZE

The herdsman, herd owner, and judge can now put the above information together and reach a decision relative to preference, but the following guidelines are offered:

1. Stature is important, but some of the deficiencies described should be avoided.
2. Size is now less controversial and many prefer a functional, medium-sized or moderately-large upstanding type cow.
3. For show-ring judging, it is important to remember that size is not everything and that cows can be too large. However, if a larger cow is of equal quality and conformation to a medium-sized cow, the larger cow will be placed up. On the other hand, a smaller cow places over a larger one that has a major defect or that can be criticized on a number of points, especially on udder.
4. Cows with exceptionally high production records and high lifetime producers are usually large, strong-framed cows with dairy quality.

SUPERIOR GENERAL APPEARANCE

Proper constitution and size for the breed, with style, symmetry, and balance, including an impressive carriage, all contribute to a superior general appearance. Cows should be representative of the breed, display signs of productive capacity, and show strength along with refinement and dairy character. Bulls should have the same superior characteristics, but they should have heavier, stronger frames.

Substance and strength refer to the general frame and size of the bone of an individual. Constitution refers to chest and barrel development.

Style is indicative of physical fitness and includes development, symmetry, and balance. A harmonious blending of all parts produces a unified structure displaying style and beauty and is conducive to excellent health, high milk production, and a long life.

A judge should recognize a lack of symmetry or proportion of parts. If the general outline of an individual shows some parts to be too long, too deep, too shallow, too large, or too small in proportion to other parts, it detracts from the style. Unevenness of topline, such as a very sloping rump, weak loin or chine, high pelvic arch or tail head and severely sickled rear legs, or a pendulous udder, can also detract from overall symmetry and balance. All parts of the body should be considered, but the most important are length and correctness of legs, straightness of topline, length and depth of body, shape and size of head, and neck conformation. Some judges overuse the terms *symmetry* and *balance*. It is much better to be specific in describing various conditions.

The style and beauty of a cow, as she moves with grace and agility or stands in a characteristic pose to display her conformation to its best advantage, is a natural expression of her temperament and pattern of behavior, or, as some call it, her "personality." For many years this was considered to be more important in the show ring than in the dairy herd, but it has gradually come to be accepted that the style, symmetry, and temperament of a cow are related not only to sustained health and productive power but also to resistance to injury. In general, these qualities enable the cow to fit into a smoothly functioning, high-producing dairy herd.

Figure 279 Uniformly good general appearance adds beauty to a herd and encourages a good dairy farmer to strive for still better care and management. (Courtesy Cornell University McDonald Farms, Cortland, NY)

Figure 280   This EX 96, four times All-American Aged Bull and All-Time All-American Aged Bull, excels in general appearance, which has contributed toward his winning several grand championships at the National Show. He is very upstanding, has an attractive head and a lot of breed character, shows a deep and smooth shoulder, possesses a deep body, is a good mover on his legs, and has symmetry and balance throughout. (Courtesy Gray View Farm, Union Grove, WI)

Figure 281   In contrast to the bull pictured in Fig. 280, this bull lacks general smoothness and blending of parts. He is plain-headed, is weak about the face, especially in his receding forehead, and is generally deficient in breed character.

Figure 282 A Brown Swiss bull that is almost perfect in general appearance. His breed character and strength, refinement of head, symmetrical body, smooth shoulders, straight top, and good legs all place him among the best of the breed. (Courtesy Walhalla Farms, Rexford, NY)

Figure 283 This bull is intermediate in general appearance. Although the deep body is generally acceptable with this type, the parts do not blend, especially the shoulders, which are too rough, and the rump, which is too prominent at the tail head. He is not as stylish about the head and neck as the bull pictured in Fig. 282, but he is far superior to the one shown in Fig. 284.

Figure 284   This bull is very poor in general appearance and breed character. The head is especially bad because of the pointed Roman nose and receding forehead.

Figure 285   A consistent winner of blue and purple ribbons, this All-American Bull displays outstanding breed character and general appearance. His attractive head, smooth shoulders, and fine carriage give him excellent style. (Courtesy McDonald Farms, Cortland, NY)

Figure 286 This upstanding young Brown Swiss cow displays outstanding general appearance. Note the length of bone in both her front and rear legs, the depth of shoulder, openness and length of ribs, her clean strong neck, and the attractive stylish head. All points on general appearance are positive except the slight dip in the chine; they blend into the preferred kind of dairy character and Excellent across on udder. As a 2-year-old she produced 19,000 lb M, placed Grand Champion at the Western National, and was voted All-American. She followed as an undefeated 3-year-old, was the Grand Champion at the Eastern National, at the Central National, and Reserve Grand at the Western National. She was a popular All-American three-year-old. This time she produced over 20,000 lb M and over 800 lb F. (Courtesy Bridge View, Mira Loma, CA)

Figure 286A This shows the same individual as in Fig. 286 when she was an aged cow and was selected as the Grand Champion at the Central National Brown Swiss Show in Madison, WI. She is now classified Excellent across on all points and later was voted the All-American Aged Cow. The Brown Swiss Cattle Breeders Association labeled her as "Today's Model." She was in a league of her own, both as a young and aged cow. All comments attributed to her as a 3 year old were repeated at the aged-cow showing when a noted judge remarked at the National, "There has never been a question in my mind who was going to be Senior and Grand Champion when this aged cow walked into the ring." (Courtesy Forest Lawn Farm, Wausau, WI.)

Figure 287 This stylish unanimous All-American Senior Yearling Guernsey heifer is a good example of symmetry, style, strength, and quality of conformation. She is upstanding, possesses a near perfect topline with the pins slightly lower than hips, showing precisely in the position prescribed for the pins. (Courtesy John and Bonnie Ayars, Mechanicsburg, OH)

Figure 288 Two Guernsey cows whose general appearance is exceptional. This 12-year-old cow looks much younger than she is. Many cows have poor muscle tone at this age, which detracts from general appearance.

Figure 288 (continued) A standout without a show record. This kind is good in anybody's herd and gives a minimum of trouble. (Courtesy Woodacres Farm, Princeton, NJ)

Figure 289 The cows in Figs. 289 and 290 show a great deal of contrast in general appearance. This upstanding open-ribbed, clean-cut All-Time All-American Brown Swiss 3-year-old displays the ultimate in general appearance (photo reversed). (Courtesy Brown Swiss Cattle Breeders Association, Beloit, WI)

Figure 290   In contrast to Fig. 289, this cow has a plain head and a heavy neck and shoulders, shows soft in the loin, is coarse over the rump, and shows too much set in her legs.

Figure 291   This flashy All-American Heifer Calf displays a world of Ayrshire character which, combined with her many strengths and overall smoothness, give her an outstanding general appearance. (Courtesy Galney Farm, Dansville, NY)

Figure 292   A Holstein senior yearling somewhat weak in general appearance about the head and neck. Nevertheless, she is so outstanding in shoulder; depth of body; topline, especially in a long, flat, nearly level rump; and admirable legs that author Trimberger selected her to head a class of outstanding heifers at the International Show. At a later date she was selected All-American. (Courtesy Carnation Farms, Seattle, WA)

Figure 293   This attractive and famous 4E 97 GMD cow excelled as a producer with four records over 1000 lb fat. Her best record at 7 years, 365d, 2X, was 26,470 lb Milk, 4.4%, 1166 lb fat. She died at 16 years of age and had lifetime credits of 209,784 lb milk, 4.5%, 9471 lb fat. She was the third generation of 200,000-lb producers. She has 10 Excellent daughters. She was a member of the All-American Produce for three successive years. She and her Ex 96 sister were voted All-Time All-American Produce of Dam. She excelled in general appearance, whether observed in the show ring or in the milking herd. It is said of this cow that "Production was her game but type brought her fame." (Courtesy Bob and Kay Miller and family, Mil-R-Mor, Dundee, IL)

Figure 294   This 2E 97 three-time All-American is the ultimate in general appearance. She was voted the All-Time All-American 2-year-old, the All-Time All-American 3-year-old and Reserve All-Time All-American 4-year-old. Her general appearance helped her in these unprecedented achievements. She is the glamorous kind. At 5 yr, 9 mo she produced 27,320 lb M, 3.5%, 956 lb F and her CTPI is +488. (Courtesy Bower Farms, Inc., Ray Vail, Manager, LaGrangeville, NY)

14
REASONS FOR PLACING
OF CLASSES

A good judge needs both a thorough knowledge of the ideal type and the ability to give effective and accurate reasons for his or her placings. Only when the judge has developed this ability can he or she teach others how to judge. At a show the judge can be very convincing and constructive with a few concise, well-organized remarks about the placing of a class.

In a previous chapter it was stated that judging involves keen powers of observation and the ability to retain what one sees in order to make a sound decision and a satisfactory judgment. By the time a judge reaches the final decision, he or she should also be well fortified with reasons to justify the placings.

The ability to give a well-organized set of oral or written reasons is extremely important. The exhibitors and spectators at a show and the students interested in judging are usually satisfied after they hear logical reasons for the placings and know that the points of conformation have been accurately observed by the official. Some placings may be controversial. However, most people realize that in matters of judgment there are bound to be some disagreements and no one doubts the honesty of the decisions. Frequently the judges of the U.S. Supreme Court do not agree and render a split decision. Sometimes there is so much difference in their viewpoints that a minority report is requested. If a dairy cattle judge has to defend a minority decision on placings with which the exhibitors and spectators do not agree, there is usually considerably more general satisfaction after the judge explains the class and gives good reasons for the placings. A noted judge at a show once used this good expression: "Intelligent reasons can be given for placing this pair either way."

The final placing of a class should depend on observing, comparing, and evaluating the differences among the various cows. Manufacturing reasons to justify an erroneous or haphazard placing is a decided mistake. Almost all students, and even experienced judges, find it difficult to give satisfactory reasons until they have had extensive training or experience. However, once the mind has been trained to absorb and retain a mental image of the good and bad points of many cows, and once a person has developed the power to recall and the vocabulary to state these accurately, concisely, and effectively, it becomes a pleasure to include reasons with judging.

BASIC REQUIREMENTS FOR GOOD REASONS

1. Know in complete detail the terminology for all parts of a dairy cow, heifer, or bull.
2. Possess a thorough knowledge of the ideal type. Know the type standard and have a crystal-clear image of each trait and overall conformation.
3. Know the specific characteristics of all breeds.
4. Develop accuracy in evaluation by working hard to sharpen your powers of observation.
5. Develop the ability to make comparisons among animals and with the ideal.
6. Build confidence to make sound decisions and good judgments.
7. Learn to concentrate on your job and to ignore distractions and interference; this will help you develop self-reliance.
8. Learn to organize the points you observe so that you can give your reasons briefly and concisely.
9. Practice speaking into a microphone until you can speak easily to a group of people without being nervous and without hesitation. Record and study your reasons. Then make necessary improvements.

Important Points For Giving Good Reasons

1. Make sure that placings are based on sound evaluation and judgment.
2. Remember that the accuracy of your reasons is essential and that your reasons must be precise.
3. Create a favorable impression by your appearance, manner, and sincerity. Relax and feel confident in your placing and statements, but do not be arrogant. Keep in mind that self-assurance inspires confidence in your judgments.
4. Emphasize correctly the functional importance of individual good or bad characteristics and indicate the degree of deviation and difference.
5. Make specific accurate statements. Use comparisons extensively.
6. List the major differences (advantages) first. Then follow with minor dif-

ferences or differences in characteristics of minor importance. Omit questionable points.

7. Grant points of superiority in the lower placing animal of a pair, and explain that these points are yielded in making the decision.
8. Use unique and colorful expressions to indicate individuality and appeal. Do not use slang.
9. Keep in mind that good grammar, complete sentences, accepted terminology, accuracy, persuasion, and pleasant, convincing, and conversational delivery are all indispensable to an outstanding set of reasons.

SCORE CARD AND GUIDELINES FOR ORAL REASONS

At the coaches' meeting held in conjunction with the National Intercollegiate Dairy Cattle Judging Contest, the group adopted a Score Card for Grading Oral Reasons and Guidelines for Oral Reasons which are to be handed to the judges before each contest. These should prove useful for good results.

Score Card for Grading Oral Reasons

A. Content (accuracy and completeness) 35
   * Accurately compares each pair of animals
   * Presents important differences first (those that clearly influenced the decision) in an organized manner
   * Uses a positive approach with comparative terminology. Grants the lower-placed animal whenever there is such an advantage
   * Briefly evaluates the bottom animal for strengths and weaknesses
B. Delivery...15
   * Uses proper terminology, grammar, and pronunciation
   * Speaks convincingly, with moderate speed
   * Is sincere, courteous, and completes reasons in allotted two minutes
   TOTAL 50

Guidelines for Officials Taking Oral Reasons

1. Avoid fraternizing with coaches and contestants prior to and during the contest.
2. Maintain a friendly attitude that will be encouraging to the contestant.
3. Ask questions only to clarify a point of misunderstanding or to determine if the contestant really saw the class. Avoid trivial questions.
4. Avoid gestures which indicate disagreement with what the contestant has stated.
5. Avoid mannerisms which may disturb the contestant.

6. Allow a contestant to develop his own individual style in presenting an effective set of reasons.

7. There is no required closing for reasons. A statement about the bottom cow is as acceptable as, "Therefore, I placed this class of Holstein cows," and so on.

8. There is no credit or penalty for saying "Thank you. Are there any questions?" This may be part of the contestant's own individual style.

9. If a contestant receives a low-placing score but says all the right things with proper emphasis, then he or she should be graded accordingly versus not getting a score above the placing score.

10. Officials should avoid discussions of classes in the presence of contestants.

RECOMMENDED METHOD OF RECORDING REASONS

Since a judge at a show cannot take written notes, he or she must take mental notes. Students in competitive judging are often required to give reasons for placing classes several hours after they have placed a number of different classes. Under these circumstances, it is a distinct advantage for the student to take notes that will help him or her remember each animal in each class.

Usually, each student develops his or her own system of recording observations in the shortest time possible. Many students use symbols similar to those used in shorthand. Good and bad points should be noted, including such features as color, a high tail setting, crooked legs, weak pasterns, deficient heart, outstanding dairy character, unusually good udder, an almost perfect rump, great style and character about the head.

Although many different systems are used to record and organize reasons for placing a particular class, one of the most useful methods is to write a brief description of each animal, because it will help one recall a mental image of the various animals later. As this is done, the placing of the class is often clearly indicated. Specific points of advantage and points that have to be granted in each pair can then be recorded.

The first requirement of a good set of reasons is accuracy in placing and in statements. Every single characteristic of the animal must be correctly evaluated, both individually and in comparison with others in the class. Correctness of emphasis is next in importance. Definite and accurate statements should be made on the points of superiority. Only those that are specific should be listed. A good judge prefers to list the points of major differences first rather than take the points for discussion in a set order. Minor differences or differences in characteristics of minor importance are considered next. Points for which the advantages are so slight that they may be questionable should be omitted. A good judge always lists the points of the second-ranking animal which are superior to those of the winner, especially in close placings, since it indicates that the points were accurately observed but were evaluated or weighted differently.

Accuracy in reasons for placing is so important that the beginner should not attempt reasons until he or she has attained some proficiency in recognizing and evaluating the ideal type. Otherwise, the beginner may develop the habit of giving reasons lacking in precision. A good set of reasons should fit a particular class so well that the reasons distinguish that class from a hundred other classes.

Reasons that are specific have a great advantage over those that are general. For example, to say that one excels over the other in body capacity is too general. The points of difference may include length, depth, spring of rib, or many other features of conformation. But to say that one cow has a definite advantage in spring of rib and a great advantage in depth as indicated by length of fore and rear rib and then point out that the other cow has a very short rib and shallow body describes these characteristics so vividly that listeners cannot help but obtain a sharp impression of these differences and their evaluation.

The use of appropriate terms, as discussed previously, is very important. The vocabulary should be broad enough to avoid monotonous repetition of words or phrases and too much use of such words as *good, better,* and *best.*

Decisive, positive, descriptive terms, such as *extremely impressive, nearly perfect, full of quality, true dairy type, a deep-bodied cow with plenty of substance combined with refinement, outstanding balance throughout, with a fine udder,* and *approaching the ideal,* are so favorable in their impression that they will be well received. To these can be added expressions frequently used by outstanding dairy people and terms fitting specific situations, such as "an old cow carrying her years lightly," "a small cow a trifle too frail for the breed," "this class was strong indeed, with three possible winners, and not a bad one in the class," "a sweet dairy heifer full of quality," "the right kind of topline," "very impressive on the move but settles a bit in the back when she comes to rest." All of these are unique, fit a specific situation, and are so colorful that they have appeal and possess individuality.

COMPARATIVE TERMS USEFUL IN ORAL REASONS

The proper use of terms to compare the differences among animals precisely is an essential part of a good set of reasons. Some comparative terminology statements are included here to help persons develop a useful vocabulary in giving oral reasons. Other descriptive terms pertaining to both desirable and undesirable conditions are listed in Chapters 6, 7, 8, 9, 10, 11, and 12.

General Appearance

I. *Overall:*
1. More overall style, balance, and symmetry.
2. More harmonious blending of body parts.
3. A larger, more upstanding, stretchier cow (bull).

    4. A taller, longer cow (bull) with more vigor, strength, and substance.

    5. A more powerful cow (bull) exhibiting more size, scale, and substance.

II. *Breed Characteristics:*

    1. More breed character (style) (alertness) about the head.

    2. Cleaner cut, more feminine (refined) head.

    3. Broader muzzle.

    4. Stronger, deeper jaw.

    5. Brighter, more prominent eye.

    6. Wider forehead.

    7. Less prominent bridge of nose (less of a Roman nose).

    8. Larger, more open nostrils.

    9. Longer and leaner (cleaner) neck.

   10. Cleaner in the throat (dewlap).

   11. Less ewe-necked.

   12. Neck blends more smoothly with the shoulders.

III. *Stature:*

    1. Taller cow.

    2. Taller and longer cow.

    3. Has more size and scale.

    4. Stretchier cow.

    5. More size for her age.

    6. A larger cow.

IV. *Shoulders:*

    1. Smoother at the top (blades) (elbows) of her shoulders.

    2. Less coarse and open over the withers.

    3. Tighter through the shoulders.

    4. Neater through the shoulders.

    5. More neatly laid-in at the shoulders.

    6. Less prominent at the point of shoulders (elbows).

    7. Cleaner (less coarse) about her shoulders.

    8. Tighter elbows (less winged shoulders).

    9. Blends more smoothly from shoulder to crops (fore rib).

   10. Fuller in the crops.

   11. Deeper in her shoulder.

V. *Back or Topline:*
1. Straighter over the topline.
2. Stronger loin (chine) (back).
3. Smoother topline.
4. Less roached loin. ◟
5. Wider loin.
6. Flatter loin.
7. Harder loin.
8. Stronger chine.
9. Longer (wider) rump.
10. Longer from hips to pins.
11. Wider hips (pins).
12. More nearly level from hips to pins.
13. Higher (wider) thurls.
14. Smoother tail head.
15. Less coarse tail head.
16. More neatly laid-in at the tail head.
17. Less prominent at the tail head (pelvic arch).
18. Roomier in the pelvic region.
19. Cleaner rump.
20. More prominent about the hips and pins.

VI. *Legs and Feet:*
1. Straighter front legs.
2. Less toe-out in the front legs.
3. Less buck-kneed in the front legs.
4. Less knock-kneed in the front legs.
5. Less sickle hocked, not as close at the hock.
6. Straighter (more correct) in the rear legs when viewed from the side (from the rear).
7. Less toe-out in the rear legs.
8. More correct angle to the hock (less post-legged).
9. Stands more squarely on her (his) legs.
10. Carries the rear legs more squarely under the body rather than standing too far back on the rear legs.
11. Moves more easily (freely) (correctly) on her (his) legs (rear legs) (front legs).

12. Stronger pasterns.
13. Shorter pasterns.
14. Deeper heel.
15. Steeper angle of the hooves.
16. Less spread toes.
17. Cleaner bone.
18. More refined, less coarse bone.
19. More substance of bone.
20. Cleaner bone.
21. Flatter bone.
22. Less puffy hock (pasterns).
23. Wider hock.
24. Less tender on her (his) feet.
25. Sounder legs.

Dairy Character

I. *Overall:*
1. Exhibits more milkiness and is sharper, leaner, and cleaner throughout.
2. Excels in dairyness and is a sharper, cleaner cut, more angular cow.
3. Is sharper, cleaner cut, and more angular.
4. Appears to be a milkier cow with more openness.
5. Displays more angularity and openness.
6. Sharper and cleaner cut throughout.
7. Carrying less flesh throughout (freer of excess flesh).
8. A more refined individual.
9. A less coarse individual.

II. *Specific:*
1. Cleaner cut and more feminine about the head.
2. Cleaner cut about the head and neck.
3. Longer and leaner neck.
4. Longer and thinner neck.
5. Cleaner throat.
6. Trimmer in the throat (dewlap) (brisket).
7. Sharper over the withers.
8. Sharper over the withers and topline.
9. Cleaner over the topline.

10. More prominent vertebrae.
11. More prominent about the hips (hooks) and pins.
12. Less fleshing over the topline, hips, and pins.
13. Cleaner through the rump.
14. Neater (more refined) in the tail head.
15. Flatter (cleaner) (thinner) in the thighs.
16. More incurving (less full) in the thighs.
17. More open ribbed.
18. More angular and open ribbed.
19. More openness of rib.
20. More open in her ribbing.
21. Cleaner boned.
22. More refined in her (his) bone.
23. Flatter, cleaner bone.
24. Less coarseness of bone.

Body Capacity

I. *Overall:*
1. Deeper, wider, longer bodied cow.
2. Larger cow with more width and depth of heart and barrel.
3. More powerful front end that is wider and deeper in the chest.
4. Longer, deeper, wider barrel.
5. A larger, more capacious barrel.
6. Deeper in both fore and rear rib.

II. *Specific:*
1. Wider (deeper) chest.
2. Wider (deeper) heart.
3. Fuller heart.
4. Stronger through the front end.
5. More width of chest floor.
6. Deeper fore ribs.
7. More spring of fore ribs.
8. Larger heart girth.
9. Fuller in the crops.
10. Fuller behind the elbow.
11. Less pinched in the heart.
12. More length of body (longer body) (stretchier body).

13. More depth of body (deeper body).
14. More width of body (spring of rib) (arch of rib).
15. Great spring of ribs.
16. More arch to her ribbing.
17. Less slab-sided.
18. Deeper flank (more depth of flank).
19. More depth of rear rib (deeper rear ribbing).
20. Longer ribbed animal that is deeper in body.
21. More width of loin (wider loin).

Mammary System

I. *Overall:*

1. More capacious udder.
2. Less pendulous udder.
3. Udder carried higher above the hocks.
4. Floor of her udder carried higher above the hocks.
5. Milkier udder.
6. More apparent quality of udder.
7. Less congestion in her udder.
8. Softer, more pliable udder.
9. More udder veining.

II. *Shape:*

1. More symmetrically shaped udder (more symmetrical udder).
2. More balance of udder.
3. More balance and symmetry of udder (more shapely udder).
4. Fuller rear quarters.
5. More correct in shape of rear udder.
6. More desirable contour of rear udder.
7. More uniform width of udder (rear udder) from front to rear.
8. Less pear-shaped in rear udder.
9. Longer udder.
10. Longer fore udder.
11. Less steep (abrupt) fore udder.
12. Less quartered (halved) (cleft) udder.
13. More balanced between fore quarters.
14. More balance of rear quarters.
15. Flatter on the sides of her fore udder.

16. Less bulgy on the sides (floor) of her fore udder.
17. Less bulgy fore udder.
18. Less bulgy rear udder.
19. More defined halving of udder.
20. More level udder floor (less tilt of the udder floor).

III. *Attachments:*
1. Stronger medial support (medial suspensory ligament).
2. Udder carried higher above the hocks (carries her udder higher) (less depth of udder).
3. Less pendulous udder.
4. Higher (wider) (smoother) (stronger) (firmer) rear udder attachment.
5. Stronger (smoother) (tighter) (firmer) fore udder attachment.
6. More snugly attached fore (rear) udder.

IV. *Teats:*
1. Squarer teat placement.
2. More correct teat (front teat) (rear teat) placement.
3. Front teats closer together.
4. Less width between front teats.
5. Teats placed more nearly plumb (more perpendicular to the ground).
6. Less strutting front (rear) teats.
7. More space between front and rear teats.
8. More desirable sized teats.
9. Smaller (shorter) teats.
10. More desirable shaped teats.
11. Less funnel-shaped teats.
12. Less bulbous (bottle-shaped) (pencil) (long) (tapering) teats.
13. Rear teats placed farther forward.

EXAMPLES OF REASONS

To give a concrete illustration for taking notes and developing a set of reasons, four cows in a Guernsey class of aged cows (Fig. 295) are used.

The first step is to write a brief description of each cow as the class is viewed. The descriptions might read somewhat like this:

No. 1: Smallest cow in class.
Prominent and rough at point of shoulder.
Shallow body, especially deficient heart behind the forearm.
Least depth of any cow in class.

Strong, straight loin, good top, but prominent tail head.
Poorest udder in class. Short fore, quartered and tilted.
    Pinched, poorly attached rear.
Excellent legs. Hornless, but stub horn on right side.

No. 2:   Outstanding and attractive in type throughout.
Excellent Guernsey breed and dairy character.
Sharp refined body with open and deepest ribs in class.
Udder attachments, shape, quality, and body veining best in class.
Nearly perfect rear udder attachment.
Strong, flat leg bone with proper set.
Slight dip in loin, but long, nearly level rump easily best of group.
Most white in rear of body.

Figure 295   Class of aged Guernsey cows. (Courtesy American Guernsey Cattle Club, Peterborough, NH)

1

2

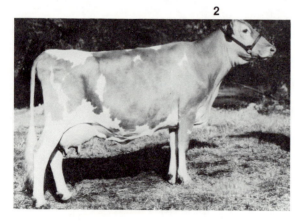

3

4

Figure 295 (continued)

No. 3:   Strong cow, lacking in dairy refinement, especially about head, neck, shoulders, and rump.

Prominent tail head, wry to right.

Weakest rear legs in class.

Poorly attached rear udder, hangs too far forward.

Excellent spring of rib.

Wide band of white from withers to forearm.

No. 4:   Largest udder in class. Good attachments. Light on left side of udder.

Deep body. Very good head but bold neck.

Trifle awkward on her hind legs.

Strong chine, slight wave in loin, small niche in rump, most prominent and coarsest tail head in class. Patchy at pins but level from hips to pins.

USE OF SYMBOLS AND ABBREVIATIONS

Much time can be saved by using symbols and abbreviations in taking notes. Usually each individual has his or her own particular system.

Suggestions for abbreviated notes on Cows No. 1 and No. 2 above are as follows:

Cow No. 1: Sma st of 4, prom & rgh pt sh, shal body def hrt esp ←f-arm, < dpt in cl, strg st ln, gd top prom t.h., pst ud in cl, short fore, ct up & titd ud, V rear ud, ex leg, no h stb r.s.

Cow No. 2: Outst & attr type, flashy—ex g br & d ch, shrp & ref body—open & dpst in class, ud att, sh, q, and body v > in class, nr prf r att ud, str fl leg b—prop set, sl dp in ln but 1, nr fl rmp > in class, > wh in rear pt of bd.

POINTS OF COMPARATIVE ADVANTAGE

After the preliminary notes have been taken, the correct placing is usually evident and the points of comparative advantage can be listed. These should be listed from most to least important. These are most useful for giving accurate and effective reasons to justify the placing. To save time, symbols and abbreviations can again be employed. However, for clarity to the reader, the following advantages for the placing 2-4-3-1 are given in detail without abbreviation.

$\frac{2}{4}$ Size, dairy and breed character, depth of fore and rear rib, openness of rib, flatness and fullness of rump, especially refinement at tail head and levelness on top. Great advantage in width, height, and strength of rear udder attachment. Balance and quality of udder. Superior body veining and set to legs. Sharpness, clean-cutness and refinement throughout. Grant strength and balance of head, especially shape of nose and forehead, cleaner under the throat, size of udder.

$\frac{4}{3}$ Great advantage in udder attachments and teat placing from side view. More capacious udder. Breed character and refinement about the head. Levelness and fullness from hip to pins. Slight advantage on set to legs and sharpness of withers. Grant strength of top, especially loin, depth of heart and smoothness at point of shoulder.

$\frac{3}{1}$ Size, strength, and depth throughout, spring of rib, smoothness of shoulder. Length and smoothness of fore udder attachment. Fullness of fore udder and levelness of udder floor. Grant great advantage on strength of hind legs. Dairy and breed character of head, and refinement of neck. Sharpness of withers and dairy character throughout.

COMPLETE SET OF REASONS

From the two sets of notes described above in this chapter, it is now possible to make up a set of reasons that specifically fits this class and every individual in the class. For the set of reasons, the senior author drew directly on his coaching experience with seven winning teams at the National Intercollegiate Contest. The same procedures were followed as those used by Cornell students who placed first for three consecutive years and during another period either high individual or high team in oral reasons for four consecutive years at the National Contest. The second author's reasons follow the senior author's reasons on the Guernsey class. Similar procedures were used by Virginia Polytechnic Institute students who were high team twice and second high team twice in oral reasons at the National Contest over a period of 6 years. Two of these team members were high individuals and another tied for high individual in oral reasons. These will be followed by students' reasons. Since almost 100, and in recent years up to 150, outstanding college students compete in the contest and the reasons are taken by five outstanding judges, it is a critical test for effective reasons. Constantly repeating reasons until they can be delivered fluently and with extreme precision is suggested.

Reasons for Placings of Guernsey Aged Cow Class
George W. Trimberger

In placing this class of aged Guernsey cows 2-4-3-1, I find an easy top in 2, a definite bottom place in 1, and an easy placing between 4 and 3. The large deep-bodied, open-ribbed 2 cow stands out in this class. Her sharpness, refinement, and outstanding dairy and breed character, combined with the best udder and rump in the class, make her an ideal top.

Two easily places over 4 because she has a great advantage in depth of fore and rear rib and in openness of rib. She is far superior in dairy character, as indicated by refinement about the neck, sharpness of withers and hooks, and especially in cleanness about the pins; four is criticized severely for patchiness over the pins. Two also excels 4 on mammary system, especially in her great advantage in width, height, and strength of rear attachment and also in balance of udder, since 4 is unbalanced and light on the left side. Although 4 must be granted an advantage for having a more capacious udder, the fore udder on 2 is smoother and stronger in attachment. Two also has quality of udder and is superior in body veining; she is remarkable in this respect.

On the topline, both cows are strong in the chine, a trifle easy in the loin, but 2 is much superior over the rump. She is much fuller over the rump and much wider at the pins; she is much more refined about the tail head, in marked contrast to the prominent tail head of 4 and the slight niche on the top of the rump, which gives 2 considerable advantage in the pelvic region. Two has the second best legs in the class and has an advantage in set of hock and strength of pasterns over 4.

In addition to capacity of udder, previously granted, it should be mentioned that 4 must be granted an advantage in balance of head, especially in shape of nose and forehead, in which 4 is outstanding; two is somewhat plain about the nose and a trifle receding in the forehead. Four is also cleaner under the throat.

Four places easily over 3. She has a decided advantage in her more capacious udder, which is much longer and stronger in fore udder attachment. From the side view, the teat placing on 4 is much preferred, since the fore and rear teat spacing is much wider and the rear teats do not point forward as they do on 3. Both cows lack good rear attachments, but 4 has a decided advantage in breed character and dairy refinement about the head. Four is the best in the class on head, but 3 is short and heavy-headed and lacks refinement. Both cows are faulty in their hind legs, but 4 has a slight advantage through the hocks and stands on a stronger bone, especially about the pasterns. Four has levelness and fullness of rump; three falls away on the side of the rump with low and narrow pin bones. Both cows are prominent at the tail head but 3 is also wry to the right.

In making the placing, 3 must be granted an advantage over 4 in spring of rib, depth of heart, smoothness at the point of the shoulder, and strength of top, especially through the loin; also in balance of udder, since 4 is light on the left side.

Three places over 1, who is a definite bottom because of her inferior udder, but each cow has a number of points of advantage. Three has a great advantage in size and strength of body, as indicated by spring of rib, depth of fore and rear rib. Although the fore udder on 3 is far from ideal, it has a great advantage over the short, quartered and deficient fore udder on 1, which makes 3 superior in levelness of udder floor and fore attachment.

Three also has a smoother and more firmly attached shoulder, but 1 must be granted sharpness of withers, dairy character, refinement, and angularity throughout. These advantages are especially evident about the head and neck, in which 3 is coarse. Since 1 has an excellent set of legs, which are the best in the class, she must be granted a decided advantage on this point in which 3 is the poorest in the class. Three is awkward on her legs, is slightly sickle in shape of hock, and has long weak pasterns.

One places at the bottom of the class, primarily because her inferior udder is the poorest in the class. She is the smallest cow with the least depth, and is the roughest at the point of the shoulder. However, she stands on the best set of legs in the class and is sharp, angular, and refined, with considerable dairy character throughout.

Reasons for Placing of Guernsey Aged Cow Class
William M. Etgen

I placed this class of Guernsey cows 2–4–3–1. Two's definite advantage in body, dairy, and udder make her an easy winner in the class. She is deeper in both fore and rear rib and has more spring of rib than 4. Two is cleaner cut and more angular throughout. She is cleaner over the topline, more prominent about the vertebrae, hips, and pins and has thinner thighs than 4. She is also more open ribbed and more refined in her bone. Two is less pear-shaped in her rear udder, is wider and smoother in her rear udder attachment, is less bulgy in her fore udder, shows more apparent udder quality, and is closer in her front teat placement. In addition, 2 is less coarse over the tail head, wider in the pins, and stronger in her pasterns. I grant 4 has more capacity of udder, is a bit stronger in her loin, has more breed character about the head, and is less throaty than 2.

Four places over 3 because she has a more capacious udder that has a longer, fuller fore udder, a fuller rear udder, and wider spacing between fore and rear teats. She also has more breed character and refinement about the head and a longer, cleaner neck. Four has wider pins and is more correct in her tail head; 3's tail head is wry to the right. I concede that 3 is smoother in the shoulders, has more spring of rib, and is stronger over the loin.

I placed 3 over 1 primarily on her advantage in strength, power, and udder. She is deeper and wider in her heart and has more depth of barrel. Three is less tilted on the floor of her udder, smoother and stronger in her fore udder attachment, and has teats that hang more plumb. She is also smoother at the point of the shoulder. I admit that 1 is a sharper, cleaner cow throughout and stands more correctly on her rear legs and pasterns. However, I placed her last because she lacks depth of heart and rib and correctness in shape and attachment of udder.

High Individual (Tie)[1] in Oral Reasons at the National Contest
Steve Ziemba, Cornell University

In placing this class of Jersey aged cows 4-1-2-3, I found a definite top in the well-balanced, very dairy, stylish, and best-uddered 4; a logical second placing in the upstanding, powerful 1 over the smaller 2, and a definite bottom in the aging, wing-shouldered, weak-uddered 3.

Four places over 1 with an advantage in upstandingness and dairy character, being a sharper, cleaner cow throughout. She is particularly sharper over those withers, hooks, and pins and cleaner over her ribs. She has the best udder in the class and has the advantage of a longer udder. She is smoother in fore udder attachment and has the advantage in balance of udder, 1 being light on the left. She also has an advantage in depth of body and strength throughout that front end, being particularly smoother and tighter in the shoulders; I criticize 1 for weakening in the point of the shoulder. Four also has an advantage in Jersey breed character throughout but particularly in the head. In making this

---

[1]This was the last set of reasons presented before the individual tied for first among 96 contestants on oral reasons for a set of each breed in the National Contest. These reasons were completed in 2 minutes.

Steve Ziemba, Cornell University.

placing, I do grant to 1 the advantage of strength of rear udder attachment and an advantage in standing on a stronger set of rear legs and pasterns which are the best in the class; I criticize 4 for being soft on those pasterns. One also has the advantage in strength of loin and in teat size.

One places over 2 with the advantage that she has in upstandingness, strength, and power, being a much larger cow. She also has an advantage in udder, being stronger and firmer in both her fore and rear udder attachment. She also stands on the more correct set of hind legs in the class. She has the advantage in strength of leg because she stands on more bone. In making this placing, I grant that 2 has an advantage in dairy character, being a sharper, cleaner cow; actually, she is the sharpest cow in the class. Two also has an advantage in being much smoother and tighter throughout the shoulders and more correct in her tail setting.

Two places over 3 in a very easy placing because of her tremendous advantage in dairy quality, sharpness, and strength throughout. She also has an advantage in having a more tightly attached udder. I criticize 3 for being too deep in the udder. Two also has a definite advantage in smoothness and firmness of shoulders whereas 3 is noticeably winged in the shoulders. Two also stands on a stronger, much more practical set of hind legs; I criticize 3 for weakening in the loin as she stands winged-shouldered in front and has the poorest set of hind legs in the class. It should also be mentioned that 2 has an advantage in Jersey breed character throughout but particularly in the head, even though she is wry faced. In making this placing, I do grant to 3 an advantage in body capacity and more strength and power in her body. I place 3 at the bottom of this class because she does lack in upstandingness, stretch, dairy character, and smoothness of parts throughout. She has to concede strength of hind legs and mammary to all the cows in the class, but I do admire her body capacity. For these reasons, I place this class of Jersey aged cows 4–1–2–3.

High Individual[2] in Oral Reasons and Jerseys at the National Contest
Dennis Cronkhite, Cornell University

In placing this class of Brown Swiss aged cows 1-3-4-2, I find an outstanding top in 1, the best-uddered cow in the class today, a cow that has extreme dairy quality and has that strength and power characteristic of the Brown Swiss breed. Following her with 3, another upstanding, powerful individual all the way through with a quality udder; 4 is an extreme dairy individual with a quality udder that is well attached, but today she lacks somewhat in stature. Two, the darkest colored cow in the class, places last because she lacks somewhat in quality all the way through and has the weakest attached udder in the class.

I place 1 over 3 because she is far superior in dairy character, being sharper and cleaner over the withers, hips, and pins. She is showing more openness of rib and is more incurving and cleaner in her thighs. She is longer and cleaner in her neck, showing more Brown Swiss breed character about her head and neck. In making this placing, I do grant that 3 has an advantage in strength and straightness of top line. I place 3 over 4 because she is more upstanding, showing greater size and scale with more overall strength and power. She is stronger in the front end, being wider on the chest floor and deeper in the heart, and I criticize 4 somewhat for toeing out. Three also has an advantage in body capacity, being longer bodied and deeper in both fore and rear ribs. In making this placing, I do grant that 4 has a definite advantage in sharpness and cleanness throughout.

I place 4 over 2 because she is far superior in dairy character, being sharper

Dennis Cronkhite, Cornell University

[2]These reasons were given the day before the individual placed first among 99 contestants on oral reasons for a set on each breed in the National Intercollegiate Dairy Cattle Judging Contest. These reasons were completed in 2 minutes.

and cleaner over the withers, hips, and pins. She is showing more openness of rib and is more incurving and cleaner in her thighs. She is stronger and smoother in fore udder attachment and is attached higher and wider in the rear. I criticize 2 for being somewhat loose in both fore and rear attachments. Four also shows more Brown Swiss breed character about the head and neck, whereas 2 is somewhat thick throughout the head and neck region. In making this placing, I do grant that 2 has an advantage in strength and power and upstandingness over 4. I place 2 at the bottom of the class today because she lacks the dairy quality and the strength of udder attachments to go any higher. For these reasons, I place this class 1–3–4–2.

Second High Individual[3] in Oral Reasons and All Breeds and High Individual for Brown Swiss on a Second High Team at the National Contest
Barbara Snider, Cornell University

In placing this class of Holstein aged cows 1–4–3–2, I started the class with two real dairy cows. I placed the sharpest cow in the class that is walking on the best feet and legs over another sharp cow but showing dry today. Then I followed her with 3, a tremendously deep, powerful cow but lacking in dairy character and strength of topline; I ended the class with 2, the smallest cow in the class with the poorest set of feet and legs.

    I placed 1 over 4 because she has an advantage in dairy character, being

Barbara Snider, Cornell University.

[3]Last set of reasons presented before this individual placed second among 93 contestants on reasons and second high individual in the National Intercollegiate Dairy Cattle Judging Contest. The Cornell team placed second on placings and reasons.

longer and cleaner about the head and neck, sharper over the withers, and cleaner about the hips and pins. She has a tremendous advantage in her mammary system because she shows more balance and bloom of udder today in comparison to the stale udder on 4. I criticized 4 for being low in her rear udder attachment, but I granted that she is stronger in fore udder attachment whereas 1 is somewhat loose and bulgy in her fore udder. One has the advantage of being a taller individual and she is straighter and stronger over the topline. She is far superior on her feet and legs; I criticized 4 for having too much set to her legs. I granted that 4 is higher and wider over the pins.

I placed 4 over 3 because she has a tremendous advantage in dairy character. She is much cleaner up front, sharper over the withers, and more prominent about those hips and pins. She has the advantage of feet and legs today, especially in strength of pasterns. She is much straighter and stronger in her topline, whereas 3 is soft in the loin and narrow at the pins. I granted that 3 has the advantage in size and scale, including depth and power. She also has the advantage in mammary system because she is much longer and stronger in fore udder attachment and shows more balance of udder.

Three placed over 2 on an easy placing because she has a tremendous advantage in size and power; she is just a taller, more upstanding individual with more depth of heart. She has a tremendous advantage in her mammary system. She is much longer, stronger, and smoother in her fore udder attachment. I criticized 2 for being bulgy and loose in her fore udder. Three has the advantage in feet and legs because she walks on a stronger leg with more substance of bone, with a stronger pastern, and a more correct set to her leg. I criticized 2 for having the poorest legs in the class. However, I granted that 2 has an advantage in her rear udder attachment. She also has a stronger topline and an advantage in dairy character because she is cleaner and sharper throughout. But I feel that she is the smallest cow in the class, seriously lacking in size and scale. She also lacks soundness of feet and legs. For these reasons I placed this class 1–4–3–2.

High Individual (Tie) in Oral Reasons and a Member of the Team That Placed First at the National Contest
Lois Remsburg, Virginia Polytechnic Institute

I place this class of Guernsey 3-year-olds 3–2–1–4. I found an easy top in the smooth stylish 3, the cow displaying by far the best udder in the class. Three's definite advantage in mammary places her over 2, for she is stronger and smoother in her fore udder attachment and higher, wider, and stronger in her rear attachment with more uniform width of rear udder. She has a more desirable teat placement, for 2's are quite wide in front, and has more udder veining and apparent udder quality than 2. However, 2 is cleaner and sharper throughout with more depth and spring of rib.

Two's tremendous advantage in dairy character places her over 1. She is

Lois Remsburg, Virginia Polytechnic
Institute and State University.

cleaner about the head and neck, longer in the neck, sharper over the withers and topline, cleaner at the hips, pins, and thighs, and displays more openness of rib than does 1. She is also a taller, longer cow that is straighter and stronger over the topline and smoother over the rump and tail setting. Two displays more depth of heart and rib and has more spring of rib than 1. She is more nearly level on the udder floor and displays more udder balance, for 1 is light in the left rear quarter. I grant that 1 is stronger in her fore and rear udder attachments and has a more desirable teat placement.

I place 1 over 4 because she is stronger and smoother in her fore udder attachment and wider and stronger in her rear udder attachment. One has a more uniform teat size. She has more Guernsey breed character about the head, is more nearly level from hooks to pins, and is wider at the pins than is 4. I admit that 4 has more depth and spring of rib, is more nearly level on the udder floor, and has more balance of rear quarters. I place 4 last because she lacks the straightness of lines and strength of udder attachments to go higher in this class.

High Individual in Oral Reasons and a Member of the High Team in Oral Reasons at the National Contest
Paul Feucht, Virginia Polytechnic Institute

I place this class of Holstein aged cows 1-3-4-2. I felt this class had an easy top in 1, the tall, strong, recently fresh cow, followed by a logical placing of 3 and 4 in the middle, and an easy bottom in 2, a thicker cow with a faulty udder. I placed 1 over 3 because she excels in size, scale, and power throughout. One stands taller at the withers, is longer from hips to pins, and is wider over the loin and rump. She is wider in her heart and deeper in both fore and rear rib than is 3. One displays more udder capacity and has a more correct front teat placement, and I criticize 3 for having a very wide front teat placement. I do grant, however, that 3 is more nearly level on the floor of the udder and displays more apparent quality of udder. One is carrying some edema today. Three also shows a more nearly level topline on the move, for 1 roaches a bit over the loin while on the move.

Paul Feucht, Virginia Polytechnic
Institute and State University.

I placed 3 over 4 because 3 excels in dairy character. She is cleaner and longer in her neck, sharper over the withers, and carries less flesh about her hips, pins, and thighs. She has more style and breed character about her head and blends more smoothly through her shoulders than does 4. Three also stands on a straighter set of hind legs when viewed from the side and rear. She has more correct rear feet, being less open in her toes and deeper in her heel than 4. I criticize 4 for being very faulty in her rear feet and legs. Three also has more udder quality than 4. I do grant, however, that 4 is more correct in her front teat placement.

I place 4 over 2 because she excels in mammary system, having a firmer fore udder attachment, a higher, wider rear udder attachment, and a more correct teat size; I criticize 2 for having extremely large teats. Four also is more open in her ribbing, is sharper over the shoulders, and carries less flesh about her hips and pins than does 2. She also has more width and depth of barrel. I do grant, however, that 2 has an advantage in rear feet and legs. She stands on a more correct set of hind legs when viewed from the side or rear and has a deeper heel. Two places last in this class, however, because she lacks cleanness and angularity and has the poorest mammary system in the class. For these reasons I place this class of Holstein aged cows 1–3–4–2.

High Individual in Oral Reasons and a Member
of the High Team at the National Contest
Andrew Ianni, Virginia Polytechnic Institute

In placing this class of 4-year-old Holstein cows 1–3–2–4, I found an easy top in the extremely straight, angular, and shapely uddered 1, a logical second in the youthful, smooth-blending 3, and a bottom pair, 2 and 4, both lacking in correctness of mammary system and general appearance. I place 1 over 3 because she exhibits more openness and angularity throughout. She is longer and leaner in the neck, freer of flesh across the topline and ribs, and more prominent about the hooks and pins. She also displays a more capacious mammary system, for she is higher, wider, and fuller in the rear udder, is fuller in the fore udder,

Andrew Ianni, Virgnia Polytechnic Institute and State University.

and has a more desirable front teat placement. I do grant that 3 is a taller individual displaying greater depth of heart.

I place 3 over 2 because she displays a more shapely mammary system and carries her udder higher above the hocks. She is higher and stronger in the rear udder attachment, is smoother blending and stronger in her fore udder attachment, and is less bulgy in the fore udder. In addition, she is straighter across the topline and stands on a straighter set of hind legs when viewed from the side. I criticize 2 for being roached in her loin and sickle-hocked in her rear legs. I do admit that 2 is a larger cow and displays more breed character about the head.

I placed 2 over 4 in my bottom pair because she is a taller, longer, more upstanding cow possessing greater depth of heart and width of chest floor. She is tighter at the point of shoulders and stronger in the chine and loin. Furthermore, she is higher in her rear udder, is more nearly level on the floor of her udder, and displays a more desirable teat size. I do admit that 4 stands more correctly on her rear legs when viewed from the side and is tighter and stronger in her fore udder attachment. However, I placed her last because she lacks upstandingness, smooth blending of parts through the front end, and shapeliness of udder.

Second High Individual in Both Oral Reasons and Placings and High Individual in Holsteins at the National Contest
David King, Cornell University

In placing this class of Holstein aged cows 4-2-1-3, I started the class with a definite top in the tall, very dairy, best-uddered 4; that was followed by the strong, deep-ribbed, recently fresh 2; who placed over the shorter white 1; and ended the class with the smallest, least dairy 3.

Four places over 2 because she excels in the mammary system, having a longer and smoother fore udder, more defined halving of the udder with an advantage in the medial suspensory ligament, a higher rear udder attachment, and a more symmetrical shaped rear udder. Four exhibits more dairy character by being more refined about the head; cleaner and longer in the neck; sharper

David King, Cornell University.

over the withers, hooks, and pins; and more incurving in the thighs. Furthermore, 4 has a cleaner and flatter bone whereas, in contrast, 2 is coarse boned. Four is more upstanding by being taller at the point of withers. She is straighter and smoother over the topline with a more pleasing rump. I recognize 2 is deeper and wider through the barrel.

Two places over 1 because of her definite advantage in size and scale as she is taller at the withers, longer through the barrel and from hooks to pins, deeper in the heart, wider on the chest floor, and deeper and wider in both fore and rear rib. Two also has an advantage in the mammary, displaying more symmetry and balance to the rear quarters, whereas 1 is light in the left hind quarter. Furthermore, 2 displays more breed character about the head in being wider in the muzzle and deeper in the jaw. I concede 1 is more nearly level from hooks to pins, and stands on a straighter pair of rear legs when viewed from the rear.

One places over 3 due to her extreme advantage in overall dairyness. She is cleaner in the neck and throat, sharper over the withers, more open in the rib, more prominent about the hips and pins, and cleaner in the bone. One has more strength through the front end, being deeper in the heart, wider on the chest floor, and deeper and wider in the fore rib. One has a stronger medial udder support, and teats are placed more squarely on the udder floor, especially the front teats. Furthermore, 1 is less prominent at the point of shoulders, and the neck blends more smoothly with the shoulders. I grant 3 is more balanced in the rear quarters, and stronger in the pasterns. However, I place 3 at the bottom of the class today because she lacks overall dairyness, strength of front end, and medial udder support. I admit she is correct in the set of the rear legs. These reasons justify my placing of 4–2–1–3.

PRESENTATION OF REASONS

The organization of reasons should follow some definite pattern, but individuality of expression should not be sacrificed. If the reasons become stereotyped or monotonous, the term "canned" might well be applied to them.

Reasons should be comparative in nature, with a minimum of description, except in special situations. Monotonous repetition of terms and expressions should be avoided, especially the word *better*. Such terms as *deeper, wider, stronger,* or *larger* are much more expressive than *better* in giving one feature an advantage over another.

It is essential that reasons be clear. Conciseness is very important, especially when a few short, accurate statements will handle the situation adequately. However, if condensed too much, reasons are incomplete and inadequate. The use of complete sentences, good grammar, and accepted terminology is indispensable to an excellent set of reasons. When combined with accuracy, a persuasive set of reasons, spoken in a pleasant, convincing, conversational manner, will attract attention and result in a favorable impression.

The scores assigned on reasons depend and vary according to the correct placing of the class designated for reasons. A good placing involving no more than a switch is a prerequisite to a good grade on reasons because direct statements of advantage must be made in order to have a forceful set of reasons. Built-in precision is necessary to make sweeping contrasts with extreme accuracy on evaluation for degree of differences. This is only possible with a satisfactory placing.

To present reasons well, one must develop the ability to take an accurate mental picture of each individual animal in a class and have these ready for recall when needed. To help develop this technique, a coach may ask detailed questions each time a student gives reasons. Soon he or she develops the mental process necessary to resee each individual while answering the questions correctly.

Another characteristic of a good set of reasons is colorful and unique description and comparison. This involves developing a personal trait that will serve the individual during the remainder of his or her life and is useful for outstanding performance in everyday tasks in many professions. It takes resourcefulness and self-assurance to develop a descriptive expression that is particularly suitable for the situation observed. Following are some phrases that describe well certain characteristics of dairy cows:

Lazy in the top line.

Clumsy on the move.

Awkward in motion.

Her entire body becomes lazy when standing.

An undignified carriage.

Blends much better, which is indicated by a dignified carriage.

Best in the class in this department.

Because of this limitation.

Attention is called.

With a particular advantage.

Whereas, in contrast.

She is the kind that will stay in business a long time.

Sensationally well-uddered.

The tall, sharp, strong, big-framed _____ .

Stronger in top in particular with a loin that is strong in contrast to

_____ .

Her unusually good head demonstrates the ultimate in breed perfection.

Impressive on the move but entirely unacceptable because she falls apart while standing.

The tremendous middle on this cow accentuates the deficiency in the heart, especially back of the forearm.

Remarkably good; incredibly nearly perfect; an unquestionable advantage; a decided advantage; an obvious point of advantage; more exact in leg conformation.

Three is far superior in legs with a strong bone, well-formed hock, and a strong, near-perfect pastern ending in a deep foot that is a great contrast and a direct opposite to the frail bone, sickle hock, soft pastern, and shallow heel on 2.

More exact in her overall conformation and full of quality.

The steps in giving good reasons include a precise, complete evaluation and proper organization. Comparison between two cows is very important and description must be kept to a minimum.

When you are delivering reasons, it is best to be informal and relaxed almost to the point of casual competency; talk in a firm but not loud voice. Deliver reasons in a logical and convincing way. Formality shown by using words like "Mr. Judge" or "Sir" should be avoided. It is much better to enter the room in a normal manner, stand about 8 or 10 feet away from the official judge, catch his or her eyes, and start telling your reasons in a normal tone of voice. Arms should be held along side or in back. An occasional gesture is permissible, but don't overdo.

The function of oral reasons is to develop useful traits and ability in the contestant and to prove to the judge that the contestant saw the class. Additional comments on the first- or last-placed cow at the time she is discussed may be appropriate. Mistakes are corrected on the spot by saying, "Excuse me, I wanted 3 over 4 for sharpness to indicate dairy character" or the word "correction" may be used, especially when the contestant is going back a sentence or more.

Finally, reasons need not be elaborate or flowery, but they must convey to the listener that the contestant saw and properly evaluated the differences between the individuals in the class.

For a forceful, precise set of reasons, the contestant and (or) judge must have a type standard and clear image of what he or she is looking for in both individual traits and overall conformation. Points of comparison for any particular pair are best listed in their order of importance or differences that determine placing. This gives necessary emphasis and avoids following the same routine on reasons. Stay away from "canned" reasons. The judge is interested in whether or not the contestant saw the animal from the same standpoint that a good, practical dairy farmer would. The approach should always be positive—giving the advantage first, then doing the granting. A set of reasons that begins by granting the "however points" before getting into the reasons appears apologetic.

Shorten the reasons to include only the basic facts about three or four points that determine the placing between a pair. Two minutes provide ample time in which to present the reasons on a class of four cows or heifers. Introductory descriptive remarks must be limited and brief or else they will spoil the reasons before the contestant gets into the "meat" for the placing. Always remember to give the reasons in an accurate, concise, and well-organized form.

For conciseness, it is good practice to say "4 places over 3" instead of "number 4 places over number 3," or even worse, "animal number 4 places over animal number 3."

Granting is very important, especially on close pairs, and the same approach suggested for giving an individual an advantage should be used.

REASONS IN THE SHOW RING

Once a person has mastered reasons for a competitive purpose or appreciates without this experience what is involved for reasons on placings in the show ring, it becomes easy for that person to please the audience. Audiences want to know why and how the individuals—especially those standing side by side—are different. The judge uses comparison, but a bit of description like "in contrast to the soft pasterns on 2" may be included. Short and concise reasons are an important part of every dairy show today and they should always be included to satisfy both the audience and the exhibitor.

One cow places over another usually because of one or two major differences, and it is often sufficient to mention these first. You should be specific on major differences, for example, stature, dairyness, mammary, legs and feet. Point out the major differences first; then you can elaborate. For example, you give a cow an advantage in udder and then you point out that it is on fore udder; rear udder; quality; teat size, shape, or placement; depth of udder; halving and involvement of the median suspensory ligament.

At the mike, where you should give reasons about as fast as the cows walk in front of you, it is very important to avoid talking about petty differences. Avoid descriptions; instead, use comparisons which give an advantage to the lead individual. Compare strengths and weaknesses. It is best to do this by being positive rather than negative in your statements, that is, by giving one cow an advantage instead of citing the disadvantage. This is suggested because the owner, leadsman, and (or) exhibitor prefers to hear about the strengths of the individual instead of weaknesses. Also, avoid general expressions, for example, "She excels in smoothness and overall symmetry." Be specific and make the reasons short and concise. If feet and legs are the deciding difference, tell this first and you will gain admiration from both the exhibitor and (or) leadsman and the audience. Rarely is it necessary to give reasons for more than the first six entries in a class and certainly never for more than the first ten entries except in an unusual situation in which a general comment may be justified or in a junior show. The reasons should be presented in a modest way but never should they sound unsure because this weakens the presentation, and someone at the halter will soon conclude that you were wrong. If you have confidence in yourself and know what you are doing by making selections in accordance with a clear image, your sureness will be communicated.

15

JUDGING DAIRY COWS

Comparative judging utilizes the procedures given in the previous chapters on parts of conformation. The judge should analyze each individual thoroughly and evaluate its merits and faults. The judge should make a mental record of these points and then balance them according to the values assigned to the different parts of conformation. An accurate balance of points is necessary when comparing individuals within a class to determine the relative value of points for udders, bodies, dairy character, feet and legs, shoulders, toplines, and overall general appearance.

Decisions should be prompt and sound. They are usually accurate and uniform if based on clearly defined and fixed ideals, especially if each animal is analyzed carefully before comparisons are made. So that no small detail is missed, the judge should make a careful study from a distance and close up while the animal is walking and standing. Prompt, clear decisions, supplemented by sound, concise reasons, are the basis of successful show-ring judging.

Good judging emphasizes correlation of type and lifetime production for breed improvement. Judging of cows makes the greatest contribution in this direction because the emphasis is on udders, dairy quality, with recognition of production and wearing qualities, feet and legs, body development, and the other points of conformation previously listed. In cows these characteristics can be observed in the mature or nearly mature form.

To help illustrate comparative judging, Figures 296 through 320 show both good and poor conformation in cows, as well as some cows of good type but having considerable differences in conformation. In view of the extensive discussion and illustrations of the different breeds in previous chapters, this chapter is limited to methods helpful in deciding whether an individual is good or poor in type.

The purpose of this training is to teach the dairy farmer, breeder, 4-H or FFA member, and the college student to pick up at a glance the strengths and weaknesses of a cow. For example, from a quick observation he or she should be able to decide that a cow is good in every respect except that she is very soft in the pasterns.

Because of the requirements for high production, the practical judge is more tolerant of some loosening on the attachments of the udder in the mature cow. As the udder grows, it develops more depth from the stress of high production. Thus the type trait of upstandingness becomes more important in order to have cows that are less prone to injury. Since the dairy farmer prefers an udder that is held up high, there is an emphasis on a good suspensory ligament to hold the center of the udder. Both stature and udder support are heritable and there is a considerable difference between sires. Therefore, selection is important.

From a practical standpoint, it is easy to be firm on the point that dairy character is very important because it is the type characteristic most highly associated with production. It must be remembered, however, that dairy character is defined as angularity and openness without weakness; thus angular cows that are also very narrow, weak framed, pinched in the heart, and showing narrow heads and muzzles are not to be mistaken as cows having a high degree of dairy character. Rather, they are frail cows. A frail cow is usually narrow throughout, especially at the hips and pins. To distinguish the difference in the sharpness of a cow that has outstanding dairy character, it is customary to say "sharpness with strength" or "strength with quality." This describes a cow that looks productive and has the power to stand the stress of high production. Strength of front end in relation to dairy character must be evaluated with caution in exceptionally high producers in peak lactation when dairy character is at its best. At this stage, strength and width in front should be evaluated on the basis of adequacy in order to avoid a negative correlation with production.

Good management has a beneficial influence on type. Management starts with the handling of the heifer. Early freshening is advantageous, but the heifer of normal size and the dry cow should not be fat because this is harmful. All of these factors are controlled by adequate feeding.

The emphasis is on good udders and sound legs and feet because these are of greatest importance. Conformation has much to do with feet and leg problems. A good foot has an even weight distribution and not too much weight on the toe or heel. Every herd owner considers good legs and feet an advantage, especially from an economic standpoint.

Judging Dairy Cows
Building A Line

All three authors enhanced their experience by watching popular, capable, professional judges build an attractive line in which the respective individuals had a suitable place. Next the need to have a crystal clear image of the ideal was recognized as a basic rquirement for good work. In practice, this is followed with a detailed analysis of the strengths and weaknesses of each entry and the practical importance of each. These are remembered for recall to rank the class and provide reasons to justify the placings. It is important that the top cow of the class has a good udder and outstanding dairy quality with overall correctness. Precision judging has been accomplished when the winners of the respective classes follow the same pattern when they compete for championship. Agreement from the ringside is important because many good judges and breeders watch the show. A vote of confidence on the high level achieved was provided from the voting on the All Americans for the Holstein breed this year. There were five unanimous selections. This is a tremendous advancement from the early days when many exhibitors carried some spare animals of different type to allow for the variations, discrepancies, and inconsistencies among judges.

Figure 296 Two years after the picture in Fig. 298 was taken, this outstanding cow was selected All-American Aged Cow (also All-Time All-American Aged Cow). The finished product represents the epitome of conformation that won the coveted All-American award for 5 consecutive years. She classified E with E across on udder. At 7 years, 9 months of age, 365 days, 2X, she produced 23,240 lb of milk, 4.4%, and 1033 lb of fat. She was the total performance winner at the National Show when she was Grand Champion and All-American Aged Cow for the third consecutive year. (Courtesy Meadow View, Bernard Monson, Gowrie, IA)

Figure 297 One year after the picture in Fig. 298 was taken, this cow was selected All-American 4-year-old by unanimous vote. She is also the All-Time All-American 4-year-old. At this age we see the expected development of udder, depth of body, and improved general appearance, especially in symmetry and balance. (Courtesy Meadow View, Bernard Monson, Gowrie, IA)

Figure 298 This upstanding open-ribbed cow with a superior udder was unanimously voted All-American 3-year-old. Compare with Figs. 297 and 296 to study development. (Courtesy Lyon Jerseys and Tom Lyon, Toledo, IA)

Figure 299 From seven All-American nominations, this cow was Reserve All-American for three successive years and All-American at 11 and 13 years. Her tremendous dairy character and excellent udder are obvious in this photo taken at 11 yr, 3 mo. During eleven lactations of 305 days she averaged 15,514 lb M, 4.0%, and 620 lb F. Her lifetime production totaled 171,800 lb M and 6864 lb F. She received her seventh classification at 14 yr, 9 mo with a score of 3E 94. She has six sons in AI and died at 15 yr, 1 mo. (Courtesy Mary Shank Creek, Palmyra Farm, Hagerstown, MD)

Figure 300 This All-American National Grand Champion, Ex 97, has a series of records of well over 20,000 lb of milk and represents the ultimate in Jerseys. She is one of the top lifetime producers of the Jersey breed and has seven Hall of Fame records including 26,740 lb of milk and 1186 lb of fat in 365 days at 7 yr, 5 mo and 25,138 lb of milk and 1117 lb of fat in 365 days at age 12 yr. 4 mo. Her lifetime production is over 200,000 lb of milk and 9000 lb of fat. Her USDA Cow Index is + 1580 M, + 48 F, and she has a type index of + 2.1. (Courtesy American Jersey Cattle Club, Columbus, OH, and Happy Valley Farm, Danville, KY)

Figure 301 Another All-American Jersey Cow. Note the upstandingness, the deep body, the general strength and dairy character, and the outstanding Jersey character. (Courtesy American Jersey Cattle Club, Columbus, OH)

Figure 302 A good working cow but a trifle deep in her udder and easy in her loin. She has excellent dairy character, as indicated by her long, clean neck and sharpness throughout. She also possesses tremendous strength and a sound udder that has enabled her to produce an average of 19,719 lb of milk and 905 lb of fat during her first six lactations. This performance has earned her a USDA Cow Index of + 1808 M, + 55 F, and + $211. (Courtesy Walabe Farms, Collegeville, PA)

Figure 303 A big, rugged young cow but lacking in dairy character. She shows entirely too much coarseness throughout. Note the heavy neck and brisket and the coarseness about the head and shoulder.

Figure 304 This 7E, 94.5, 3X National Grand Champion, and 5X All-American was admired for her well-attached udder carried up high by a super suspensory ligament and for her sharpness and strength that gave her outstanding dairy character. She was a quality cow that carried her age lightly. (Courtesy Woodacres Farm, Princeton, NJ)

Figure 305   A deep-bodied, strong, powerful cow with unusual refinement and Guernsey breed character about the head, neck, and shoulders and throughout her body. She was National Grand Champion and All-American. (Courtesy American Guernsey Cattle Club, Peterborough, NJ)

Figure 306   A short, dumpy cow seriously lacking in dairy character that is indicated by her round, tight ribs; coarseness about the head, neck, and shoulders; and particularly by her throatiness. This cow had very low production.

Figure 307 A 6X Very Good Cow that produced consistently (up to 17,000 lb of milk) year after year. Her rump with pins slightly lower than hips promotes good genital health. She calved at 2 yr; at 2 yr, 11 mo; at 3 yr, 11 mo; at 5 yr; at 6 yr, 1 mo; at 7 yr, 2 mo; at 9 yr, 11 mo; at 10 yr, 11 mo; and at 12 yr of age. This is a remarkable record. (Courtesy Kla-Nol Acres, Farmersville, OH)

Figure 308 This EX 93 was Grand Champion Guernsey cow at two national shows and Total Performance winner at all three national shows the same year. She was unanimous All-American Junior 2-Year-Old, All-American 3-Year-Old, and is an outstanding producer.

1y-11m	2X	365d	17,610 lb milk	4.4%	781 lb fat	
2y-11m	3X	365d	26,090 lb milk	6.7%	1737 lb fat	(World's
		MC Deviation +8,348 lb M +473 lb F				Record)
		CIT +2.6, CPI +146				

She excells in dairy character, general appearance, and has a deep, open-ribbed body and quality udder. (Courtesy Elm Spring Farm, Lynn, IN)

Figure 309 This All-American 2E across Brown Swiss was a consistent winner in the show ring and was outstanding in production with 23,230 lb of milk, 4.3%, and 1002 lb of fat at 5 yr, 10 mo. (Courtesy Leon Button, Rushville, NY)

Figure 310 This 5E Brown Swiss Cow, the former World's All-Breed lifetime butterfat producer with 13,607 lb of fat (308,569 lb of milk) in 4515 days, has ten consecutive records of over 24,000 lb of milk and 1031 lb of fat. Six of the ten records were made in 305 days with unusual regularity of calving. Note the strength of frame, the well-attached udder, including the strong medial suspensory ligament, and the overall conformation of this hard worker. (Courtesy Brown Swiss Breeders Association, Beloit, WI)

Figure 311 This cow and the cows in Figs. 312, 313, and 314 placed in order for the first four in the Aged Cow Class at the Central National Holstein Show. This cow won first, won Senior and Grand Champion, and was later voted All-American. The judge's comment about all four of these cows: beautiful, beautiful cows. One placed over 2 because she had the best udder in the class and had tremendous quality of udder. She also was given an advantage in blending of parts, tremendous shoulders, sharp withers, wide chest, full crops, and width at rump (photo reversed). (Courtesy David Bachman, Pinehurst Farms, Sheboygan Falls, WI)

Figure 312 This beautifully blended, sharp dairy, and powerful cow placed second with an advantage over 3 on dairy quality, length of body, great front end, and a very correct leg in which she has an advantage over 1. Her dairy quality (33,013 lb of milk, and 1123 lb of fat in 365 days at 7 years) combined with strength of frame and power balanced out the advantage 3 had on mammary with the second best udder in the class. She was Senior and Grand Champion at the Canadian National Exhibition, first dry cow at the Royal Winter Fair, and second at the Ontario Spring Show. (Courtesy J. M. Fraser & Sons, Spring Farm, Brampton, Ont.)

Figure 313 The cow pictured here placed third and had the second best udder in the class. She was cut from the same mold and the top 3 followed the same pattern. She excelled 4 in udder quality, attachments, and especially teat placement. Three also had considerable advantage in head and neck, depth of heart, and general strength of frame, especially in legs and feet. (Courtesy Lowell E. Peterson, Hutchinson, MN)

Figure 314 This youthful-appearing aged cow placed fourth. She was tall enough and strong enough and had a superior rear udder which placed her over a smooth and outstanding cow in fifth place (photo reversed). (Courtesy Paul Ekstein and Spring Farm, J. M. Fraser and Sons, Brampton, Ontario)

Figure 315 This Holstein cow seriously lacked in dairy character and she displayed a certain degree of coarseness throughout. She was very much lacking in productive power and, because of her weak pasterns and shallow feet, did not wear.

Figure 316 This high producer had the willingness to milk but not the strength and staying power to go along with it. Thus, like so many cows, she wore out as a relatively young cow.

Figure 317 This great 3E 97 All-Time All-American Aged Cow was named Grand Champion at the Eastern and Central National shows. She also was the Supreme Champion for all breeds at the Central National. Later, she was voted unanimous All-American Aged Cow. Concurrently she completed a record of 44,143 lb of milk and 1698 lb of fat at 5 yr, 9 mo of age. She repeated as the All-American Aged Cow the next two years and concurrently produced 48,731 lb M, 4.2%, 2028 lb F. This is the highest record of any All-American. She was also All-American 3-year-old and was voted All-Time All-American Aged Cow in the form pictured, which was during the year she won her second All-American Aged Cow Award. Her production listing was +2185 lb M, +.14%, +106 lb F, +$318 and she was also first on the index listing with a CTPI +1088. (Courtesy Woodbine Farms and Romandale Farms Ltd., Arville, PA and Unionville, Ontario, Canada)

Figure 318 This EX 97, 3E All-Time All-American Holstein Cow and later Reserve All-Time All-American made three consecutive records of over 24,000 lb of milk and over 1040 lb of fat. Her highest record was over 26,000 lb of milk at 9 years. She was an outstanding transmitter with four sons showing marked improvement in production and type as published in Holstein Sire Performance Summaries. (Courtesy Paclamar Farms, Louisville, CO)

Figure 319 This 3E 93 Cow with 29,674 lb of milk and 1230 lb of fat on 2X is outstanding in every department. Note her overall strength and quality. She is the trouble-free kind, especially on feet and legs. (Courtesy Carnation Farms, Carnation, WA)

Figure 320 This EX 97 Holstein cow shows a tremendous display of sharpness, angularity, dairy character, general appearance, with a near perfect udder and good feet and legs. She combines quality with strength and is a good worker with 20,000 lb M on 2X at 2 yr, 6 mo and 21,786 on 2X at 3 yr, 7 mo. These were followed with two outstanding records as follows:

4 yr, 11 mo	343d	3X	37,340 lb M	3.5%	1,310 lb F
5 yr, 11 mo	365d	3X	39,015 lb M	3.6%	1,400 lb F

Concurrently she was voted unanimous All-American 4-Year-Old Cow; All-Time All-American 4-Year-Old Cow; All-Canadian 4-Year-Old, 5-Year-Old, and Mature Cow. Also unanimous All-American Aged Cow two successive years, following First Aged Cow, Production Winner, Best Udder in class and of show, Grand Champion and Supreme Champion three successive years at the Central National. (Courtesy Hanover Hill Holsteins, Port Huron, Ontario, Canada, and Millerton, NY)

16
EVALUATION OF DRY COWS

It is always difficult to do justice to a dry cow in the show ring without taking too much of a chance. A judge must realize that in the average practical herd about 16 percent of the cows are dry and that even the prolific reproducer is dry 6 weeks of the year. Therefore, a dry cow in a dairy herd is a perfectly normal occurrence. However, in the past, enough mistakes have been made in selecting dry cows that it has gradually become the rule to prefer cows in milk.

Some judges compromise by giving due regard to dry cows with a previous show record, but this cannot be advocated because it conflicts with the philosophy of judging: taking the cows as they appear on the day of the show without regard to their previous appearance. In the light of this, dry cows should be properly evaluated and placed in accordance with the judgment that can be arrived at from the information at hand.

Many breeders realize that a good cow of recent freshening has the most appeal, and purposely breed cows for freshening at the time of the shows. A judge should recognize this advantage for cows that have just freshened, or nearly so, and should feel certain that the animal actually is superior before giving her preference. The judge should also recognize that many cows have won when they were in full bloom and appealing to the eye over cows that were superior when both were at the same stage of lactation.

The easiest time to evaluate a cow accurately is at mid-lactation, but a good judge, in order to make valid comparisons, must be able to judge cows at all stages of lactation and to visualize how each one would look at mid-lactation. This does not contradict the statement made above about analyzing cows the way they appear on the day of the show. It does mean, though, that a judge should select as winners cows that will be good 365 days of the year. The judge should not select the cow that is particularly good for 2 or 3 days only, but rather one whose appearance and qualities are outstanding during all stages of lactation throughout the year.

Although it is much easier to determine the udder conformation of a cow in full milk than a dry cow, a judge has many reliable indications available for evaluating the type of udder on a dry cow. Some of the telltale signs for good and bad udders are illustrated in the photographs in this chapter. Wide, firm udder attachments (up high and well forward), texture and balance of udder, and teat placement can serve as guides to evaluate the completely collapsed udder on a dry cow. Actually, in many instances, a completely collapsed udder can be more accurately evaluated than one that is beginning to fill before freshening. For example, different qualities of tissue and texture are more easily detected in a dry blind quarter.

Dry cows should carry only a moderate amount of condition. Overconditioning is a serious fault. The higher condition natural to dry cows, especially

Figure 321 Dry udder on a Guernsey cow. The high, wide, and smooth rear udder attachment, the proper length of fore udder, terminating in a smooth, strong attachment to the body, together with the correct shape and level floor of udder, and properly placed teats mark this as an excellent dry udder. (Courtesy McDonald Farms, Cortland, NY)

Figure 322 Very good dry udder, with strong fore and rear attachments and correct placing of teats. The folds in the rear udder denote quality of udder and a good attachment. The udder and body veining is outstanding. (Courtesy J. M. Fraser, Streetsville, Ontario, Canada)

over the back, should be recognized and should not be confused with thickness and coarseness, which show up more through the neck and thighs, indicating a lack of dairy quality. A careful study of the skeletal frame will reveal considerable information about openness and dairy quality on a dry cow. Additional stress at calving in the overconditioned cow often causes trouble known as the fat-cow syndrome.

With proper analysis and careful evaluation, a judge can do justice to good dry cows without taking undue chances. Dry-cow classes can be advocated for the shows because they reveal much more accurate information about udder conformation than do heifer classes. This is only a partial solution, however, since the dry classes are sometimes dominated by springing cows that are "close up" and show a fully developed udder just before freshening. Figures 323, 324, and 329 through 333, which show a dry collapsed udder as compared with a distended one on the same cow, will help in proper evaluation of the dry udder. Photographs of the entire cow during the dry period demonstrate the normal and objectionable changes in body conformation due to condition during this period.

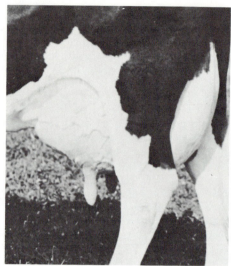

Figure 323   (A) Dry, collapsed udder. (B) Udder in milk of same cow. A good udder but the teats are too large. In each instance there is a slight discrimination for this condition.

Figure 324   Acceptable dry udder (A) reveals same faults that can be observed in milking udder (B) of this Holstein cow. Both stages show a rear udder that hangs too far forward and a fore udder that lacks strength and smoothness of attachments.

Figure 325 This outstanding show cow in California is fairly well covered and carries a trifle too much weight in the dry condition.

Figure 326 Poorly attached dry udder, with teats placed too close together. The structure for the upper region of teats is funnel-shaped. The fore udder is entirely too short and loose in its attachment (a serious discrimination).

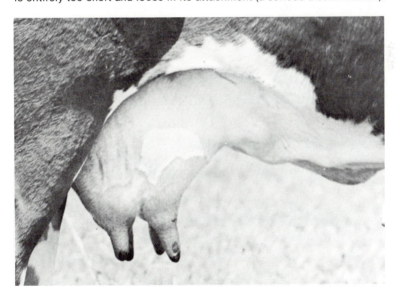

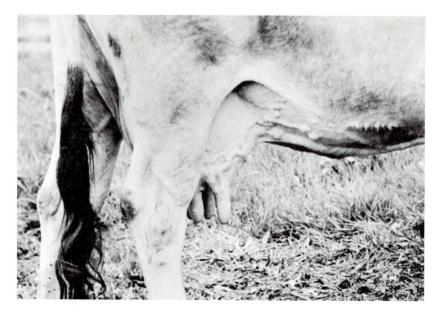

Figure 327   Dry udder with broken fore and rear attachments and funnel-shaped teats that are too close together. The udder floor is poor (a serious discrimination).

Figure 328   Pendulous dry udder with broken fore and rear attachments. It is easy to visualize the ill-shaped udder that will result when filled with tissue and milk because the medial suspensory ligament is very faulty (a serious discrimination equivalent to a disqualification).

A

Figure 329 This All-Time All-American Aged Cow and later Reserve All-Time All American Aged Cow (3E 97) was a trifle too heavy when shown dry (A) in comparison to her outstanding dairy quality (B) during high production at an advanced age. The extra weight (A) is reflected on leg posture. (Courtesy Paclamar Farms, Louisville, CO)

B

A

Figure 330   An outstanding show cow in California shows to advantage in proper dry condition (A) and in full bloom (B). (Courtesy Nunesdale Farm, Santa Rosa, CA)

B

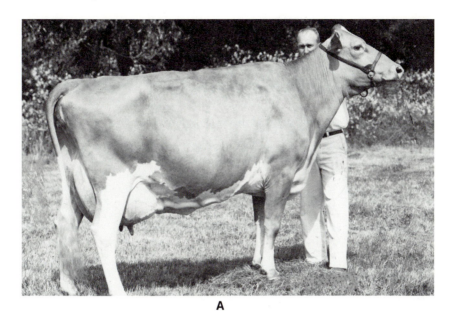

A

Figure 331 (A) Changes in body conformation during a 6-week period are shown by this Reserve All-American Guernsey with 20,000 lb of milk and 1100 lb of fat. In (B) this cow is in heavy milk flow and displays considerably more dairy character. (Courtesy Woodacres Farm, Princeton, NJ)

B

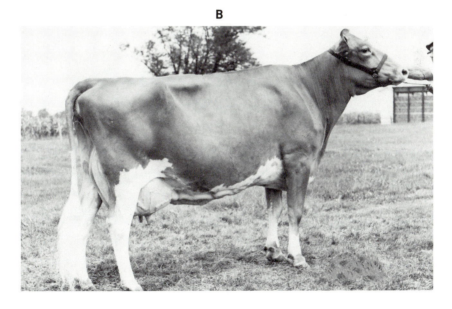

Figure 332   An outstanding cow (96E) twice Reserve All-American in milking form (Figs. 371 and 372) and All-American Yearling (Fig. 370) looks the part as a dry cow with worlds of udder quality. (Courtesy Collins-Crest Farm, Perry, NY)

Figure 333   The same cow as in Fig. 332 with well-attached udder at advanced age after many records of high production (22,909 lb of milk at 2 years and 31,028 lb of milk at 5 years). (Courtesy Collins-Crest Farm, Perry, NY)

Figure 334 Dry cow of outstanding type that has the proper amount of body fleshing for the dry period. An animal in this condition carries ample reserve for high production but has no excessive fat and body fleshing that would cause undue strain on the body system at the time of calving. The dry udder displayed by this cow shows every indication of possessing correct conformation when filled with milk. The folds in the rear udder usually indicate outstanding quality and good attachments. This cow photographed when she placed Grand Champion in the junior show and eighth in a strong open class at the International Dairy Show. (Courtesy A. C. Thompson, Elgin, IL)

17

JUDGING DAIRY BULLS

The primary purpose of a dairy bull is to produce the right kind of daughters. Selected either by a planned mating or as a young sire, the bull has to be very outstanding in his pedigree for production and type, especially in predicted differences for these characteristics as presented in the Sire Performance Summaries. Important are the following:

1. An unusually good pedigree showing ancestral superiority as indicated by a high pedigree index (PI).
2. Acceptable type traits. Freedom from weaknesses, for example, soft pasterns or shallow feet which are transmitted.
3. An expectation that the sire will prove to have unusual powers for transmitting high production or high predicted difference (PD) for milk, composition, and high wearability, especially well-attached udders and correct feet and leg structure, to his daughters.

Point 3 above is the most important, and, in the final analysis, determines the value of the contribution a sire will make toward herd and breed improvement. High production is essential from an economical standpoint, and good type in the daughters of a bull is necessary if the daughters are to have high-production records over a period of years. In addition to adding appeal and popularity to a bull, the first two points also greatly increase the odds for success with point 3.

It must be admitted that some bulls of inferior type transmit good type to the daughters, and, conversely, that some bulls of good type produce inferior daughters. But the odds are in favor of a good-type bull having daughters of good type. A herd sire having weaknesses such as poor legs, shallow body, bad topline, or weak constitution will usually pass these on to his daughters.

There should be little argument on the fact that good character, a strong and straight top with a nearly flat rump, deep ribbing for body capacity, smooth shoulders, good legs, dairy refinement, and openness of body conformation, a strong vigorous frame, breed character, with symmetry and balance throughout in the body conformation, are all inherited from the sire and dam and can be observed in both. The mammary system, or udder, although inherited through both sire and dam, must be studied through the sire's daughters, the grandsire's daughters, a sire's dam and granddam, as well as maternal sisters and brothers and their progeny. Official type classification can provide this information and will be discussed in a later chapter.

The individual type characteristics previously considered apply to bulls in the same manner that they apply to cows except that bulls should have masculinity and more strength, especially in the head, neck, and shoulders. The allotment of points on the bull score card are general appearance 55, dairy character 25, and body capacity 20.

Figures 335 through 342 illustrate good and poor conformation in a bull.

Figure 335 This outstanding EX 96 Jersey bull was five times Grand Champion at the National Jersey All-American Show and five times All-American Award winner. Note his openness of body, his strength with Jersey character, and his overall symmetry. (Courtesy Heaven Hill Farm, Lake Placid, NY)

Figure 336   This three-time All-American Holstein Bull scored 97 at 8 years, 5 months of age. He transmits his good qualities, especially legs and feet. He is an exceptionally good mover and at almost 9 years of age he moved about the ring with ease. He was voted Reserve All-Time All-American Aged Bull. (Courtesy C. M. Bottema, Jr. & Son, Indianapolis, IN)

Figure 337   This four-time All-American Aged Bull on four successive years previously was Reserve All-American 2-year-old and later All-Time All-American Aged Bull. He was also the sire of the All-American Get of Sire for two successive years. This EX 96 bull is a large, upstanding bull that is very outstanding in general appearance. His attractive head displays an abundance of breed character. Note his deep, smooth shoulder, depth of fore and rear rib, good legs, and all of these with symmetry and balance throughout. (Courtesy Gray View Farm, Union Grove, WI)

Figure 338 This outstanding Holstein sire with a classification of EX 96 points was truly an outstanding example of individual excellence and a great transmitter of both type and production. Two of his daughters and a granddaughter are the three All-Time All-Americans represented on the cover of this book. They are all classified 97, all three times or more All-American winners and have outstanding production. This bull represents a new era in dairy cattle breeding. He died at 14 years of age and has 54,843 AI tested daughters (a world record) in 10,280 herds. His daughters averaged, on ME basis 18,440 lb milk, 3.7%, 673 lb fat. The predicted difference of his daughters is +366 lb milk, +23 fat, and +$61 value. He has over 1000 registered daughters that produced over 1000 lb fat and over 200 with more than 30,000 lb of milk in one lactation. His HFAA type proof is +2.19 PDT on 40,961 classified daughters with an average age-adjusted classfication score of 83 points. He has 2862 Excellent daughters; the first sire of the breed to exceed 2000. He has over 10,000 registered sons, including many outstanding individuals. He has sired 18 individual All-American winners and three All-American Get of Sires. The All-Time All-American Aged Cow pictured on the cover is his highest daughter on milk production, with 48,731 lb milk, 4.2%, 2028 lb fat, but a California cow excells in fat production with 2104 lb fat in a year. All of this story is a great tribute and testimonial to the AI industry. (Courtesy Select Sires, Inc., Plain City, OH)

Figure 339 This All-American Senior Yearling Holstein Bull is an excellent example of size and straightness and quality. (Courtesy Mackenthun and Flemming Farms, Brownton & Cosmos, MN)

Figure 340 This All-American and Reserve All-Time All-American Junior Yearling Brown Swiss Bull is outstanding in straightness of lines, smooth blending of parts, and quality. (Courtesy Broadacres Farm, Ohio, IL)

Figure 341 This All-American Ayrshire Bull Calf is another example of size and excellent dairyness. The following year he won the Senior Yearling Bull Class and was named Junior and Grand Champion of the National Show. He also won his second All-American Award as a Senior Yearling. (Courtesy Towerview Farm, Cochranville, PA)

Figure 342 An example of an excellent Milking Shorthorn senior bull calf who earned Reserve All-American honors. (Courtesy Samuel G. Yoder, Shoemakersville, PA)

18
JUDGING DAIRY HEIFERS

Many selections are made from young animals, and a proper knowledge of the factors to consider for type characteristics at an early age can have a great deal of influence on the results obtained. For immature animals, the indications of type that receive major emphasis should be limited to those points that determine whether the individual will ultimately develop into a cow with outstanding performance over a period of years or into one with very limited capacities. During recent years considerable progress has been made toward more successful selections of heifers both in the show ring and on the farm.

Considerable study and experience are required before it is possible to select heifers successfully. Conformation changes take place as the animal matures, and these must be recognized and observed to attain proficiency in heifer selections. It is now generally agreed that the sleek, round-bodied calf or heifer does not have the odds in her favor for a good future. Such individuals will usually mature early and have insufficient stretch and frame structure to develop into a good dairy cow with the recommended scale, dairy quality, and forage-handling capacity. A trim, clean-cut, well-grown, open-ribbed heifer, with smooth shoulders, straight top, good legs, proper dairy refinement, and outstanding breed character usually develops into an outstanding dairy cow. Since proper conformation for all these type characteristics has already been discussed, it is not repeated here, but production in sire and dam is emphasized.

A clean-cut, angular heifer that is sharp and open throughout the body, indicating outstanding dairy quality, is much preferred to the short, compact individual that exhibits too much fleshing and thickness. An extremely fleshy animal is usually overfitted, with a corresponding loss of quality and refinement, and must be given serious discrimination because this kind of heifer generally has a dim future as a useful dairy cow. Experimental evidence shows that they produce less milk and have a shorter life.

It is difficult to determine the correct type in young animals and to predict their mature form, especially when it comes to the characteristics that change with age. There is no published uniform score card available to use as a guide for evaluating heifers. In regard to relative importance of the components of type in dairy heifers, general appearance should be allotted approximately 55 to 60 points, dairy character 20 to 25 points, body capacity 15 to 20 points, and mammary system about 5 points.

The major general appearance points to consider are straightness of topline, correctness of legs and feet, and size for age. A tall, long, straight-lined, smooth-blending heifer is much preferred to a more compact heifer that has faulty topline or faulty feet and legs. The head and neck should be a combination of femininity and strength and should blend the entire frame into a straight, stylish, well-grown, symmetrically shaped heifer.

Heifers and especially calves should be clean-cut and angular and should have long, clean necks, sharp toplines, open ribbing, and clean bone. Slightly more fleshing is acceptable in yearling heifers, especially senior yearling heifers as they approach calving. Thicknesss and coarseness, however, are severe criticisms in heifers of any age.

Length of body and spring of rib are desirable characteristics of heifers. Heifers with moderate depth of ribbing that is open and well sprung are very acceptable as additional depth, especially of the rear part of the barrel, usually accompanies increasing age.

Although mammary development is difficult to predict, the placing of teats is of some importance. The future size of teats can be determined fairly accurately. Widely spaced teats when viewed from the rear, very closely spaced teats when viewed from the side, or very long or strutting teats receive slight to moderate discrimination depending on the degree of faultiness. It is not possible to determine udder shape and strength of attachments in the heifer with a high degree of accuracy, except occasionally when the developing udder is very poorly carried and placed.

SEVERE DISCRIMINATION FOR OVERCONDITION IN YOUNG HEIFERS

It is extremely important in judging calves and young heifers to give considerable emphasis to their growth and development with no indications of fat or coarseness. Early fattening of dairy heifers is directly antagonistic to the purpose for which the animal will eventually be used. The fat accumulates in the udder and many other parts of the body, adversely affecting dairy type and production performance. When dairy quality and refinement are replaced by coarseness and patchiness, productive capacity is greatly reduced and body conformation is poor throughout the individual's shortened life. The resulting firm, fatty udders, with little pliable secreting tissue, and the poor udder attachments lower the level of production. These statements are not based on opinions but on results of carefully controlled feeding experiments at agricultural colleges in the United States as well as in many other countries.

Overcondition shows up in excess fleshing and patchy fat deposits, especially an extreme development of the neck, shoulders, crops, back, loin, and thighs. Deficiency in the region of the crops and lack in spring of ribs in the upper region of the fore ribs indicates underfeeding and improper nutrition and can be used as a gauge to determine the proper state of nutrition or feeding practices. If excess fleshing and overcondition do not appear until the heifer is pregnant, they are not as serious and usually disappear as the heifer comes into milk as a 2-year-old. Some extra feed is needed to get senior yearlings into moderate condition of flesh after they are pregnant. Thus some extra flesh can be tolerated in the pregnant senior yearling but never in calves or young heifers, where it must receive a very serious discrimination.

The emphasis on proper growth, without any excess flesh, in calves and heifers again demonstrates the practical application of judging procedures and how an appreciation of type can serve as a guide to improved and practical dairy cattle production.

Figure 343 A Holstein senior yearling that placed well up early in the season and then dropped to the lower part of the class later on because she became too thick. Note the heavy, short neck; the coarseness and prominence at the point of the shoulder; and the covering over the ribs, hips, and pins.

Figure 344 This smooth, upstanding, good-framed Holstein heifer is standing on exceptionally good legs. She won at two national shows and was designated Reserve All-American Junior Yearling. (Courtesy C. M. Bottema, Jr. & Son, Indianapolis, IN)

Figure 345 An All-American Junior Yearling Brown Swiss Heifer that exhibits a lot of style and breed character. She has a strong frame, ample refinement, a deep body, and good topline. (Courtesy Wayne Sliker, St. Paris, OH)

Figure 346   An outstanding Jersey Senior Yearling Heifer that displays an abundance of Jersey character about the head and throughout her body. She is a strong, upstanding heifer with ample refinement. (Courtesy Danny Weaver, Cary, IL, and Ron Buffington, Columbus, OH)

Figure 347   This All-American Senior Yearling Guernsey was a National Junior Champion. She is a strong, well-developed heifer without any sacrifice on quality. (Courtesy Woodacres Farm, Princeton, NJ)

A

Figure 348 Two All-American Ayrshire heifers that possess some differences in type. (A) This heifer is more upstanding and has a strong frame, especially in leg and pastern. (B) This heifer has the advantage of greater depth, especially through the heart region. (Courtesy Caverly Farm, Clinton, ME)

B

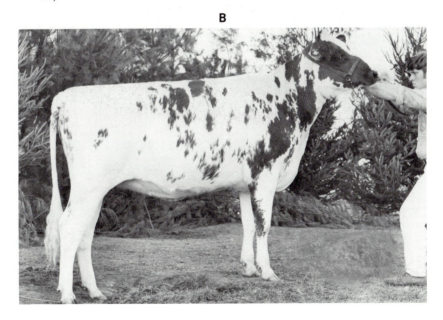

A

Figure 349 (A) Reserve All-American Junior Yearling Guernsey Heifer that improved with age and developed into an All-American 2-year-old. (B) Note the improvement in openness and dairy quality, in strength of topline, especially in loin, and in depth of ribbing. Her style and symmetry have greatly improved. (Courtesy Woodacres Farm, Princeton, NJ)

B

A

Figure 350 (A) A very outstanding All-American Senior Yearling. She is particularly pleasing in upstandingness and in dairy character. She has general strength and unusual smoothness. (B) The same individual as an outstanding 3-year-old. She is now classified Ex 93 and is an outstanding producer. Note the resemblance in overall conformation from the senior yearling to the milking stage and consider the importance of a good udder to accomplish this. (Courtesy Moncony Farms, Inc. Spencerport, NY)

B

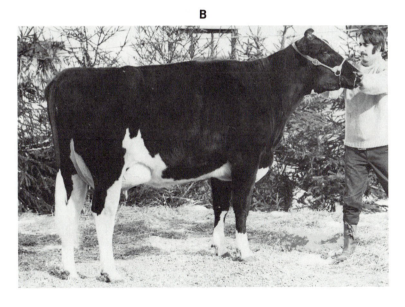

Figure 351 This Brown Swiss All-American Senior Yearling Heifer is the right kind. She was first and Junior Champion at both the Central and Eastern National Show. Note her deep, smooth shoulder; general appearance, including sound legs and feet, and her overall style and symmetry, including outstanding dairy quality providing the desirable combination of strength with quality. (Courtesy Voegeli Farms, Inc., Monticello, WI and Voegeli-Pinehurst Farm, Sheboygan Falls, WI)

Figure 352 This outstanding Guernsey senior yearling sets a pattern because she was voted unanimous All-American Senior Yearling after she was Junior Champion Female at two national Guernsey shows and Ohio State Fair. There are no weak points on this heifer. She is especially pleasing in her strong top with pins slightly lower than the hips; her quality with strength combines smoothness and power throughout. She is unusually good in cleanness, breed character, and general appearance. (Courtesy John and Bonnie Ayars, Mechnanicsburg, OH)

Figure 353 An undefeated All-American Milking Shorthorn Senior Calf. This calf is exceptional for her correctness of topline, rear legs, and dairyness. (Courtesy Michael E. Reed, Daleville, IN)

Figure 354 This All-American Milking Shorthorn Senior Yearling was undefeated over two show seasons. She was also All-American Senior Calf the previous year. She displays outstanding Milking Shorthorn breed character throughout. (Courtesy Michael E. Reed, Daleville, IN)

Figure 355   The right kind of All-American Ayrshire Junior Yearling Heifer. Note the breed character, dairy character, and smoothness throughout. (Courtesy Elms Creek Farm, Jo Ann Perdue, Omega, OK)

Figure 356   Unanimous All-American Ayrshire Calf. An upstanding, good-framed, well-developed calf, standing on exceptional legs and feet. (Courtesy Galney Farm, Dansville, NY)

Figure 357 Reserve All-American Heifer Calf from a dairy farmer's herd in Indiana is the right kind for the Holstein breed. The strength of frame without coarseness together with balance and style is outstanding. (Courtesy Baugo Valley Farm, Glen E. Cook, Elkhart, IN)

Figure 358 Senior heifer calf displaying good condition, development, and outstanding type. The depth and openness of ribbing, the smoothness of shoulders, and the long, clean-cut neck and refinement about the head combine to demonstrate fine dairy quality. (Courtesy Collins-Crest Farm, Perry, NY)

Figure 359 Holstein senior yearling that must be very severely criticized for coarseness and overcondition, a serious discrimination. Note specifically the coarse, thick head and neck; the heavy brisket; the rough point of shoulder; the lack of prominence at the hips; and the heavy, coarse tail head with untidiness about the pins. This heifer has been overfed and her productive capacity, udder quality, shape, and attachments have been ruined.

Figure 360 Attractive senior yearling properly conditioned to display outstanding dairy quality. In contrast to the heifer shown in Fig. 359, she had a successful show record and followed this with an outstanding 2-year-old record under practical farm conditions. Endurance and wearing ability are indicated by her refinement with strength about the head; the long, clean-cut neck; smoothness and depth of shoulder; tidy but deep body with an open rib, clean rump, and strong refined bones in her legs and body frame. (Courtesy Collins-Crest Farm, Perry, NY)

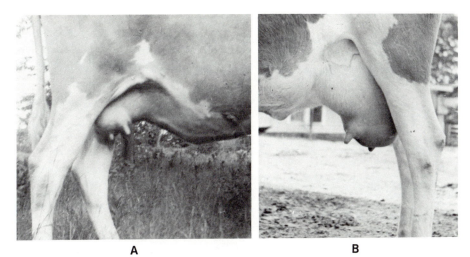

A B

Figure 361 (A) An ill-shaped udder on a heifer. The entire udder is hung too far forward. This, combined with a short fore portion, results in the whole organ being tilted. (B) The same udder with more development. The faults mentioned are more pronounced (a serious discrimination).

Figure 362 Satisfactorily shaped udder and attachments, but the teats are placed too close together and bulge at the top and approach a funnel shape at this early stage (a serious discrimination).

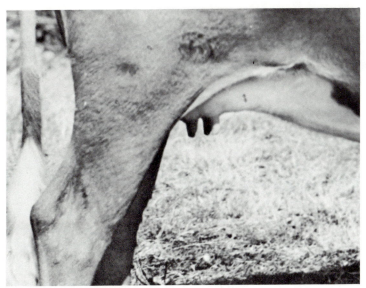

Figure 363 This heifer's udder displays a severe quartering condition and uneven teat development and too much enlargement of the fore quarter (a serious discrimination).

Figure 364 Tilted udder caused by a deficient, short fore udder and a rear udder that hangs too far forward (a serious discrimination).

Figure 365 Well-developed udder on a heifer. Note the long, strongly attached fore udder; level floor of udder; and widely spaced teats that are of correct size and properly placed on the udder.

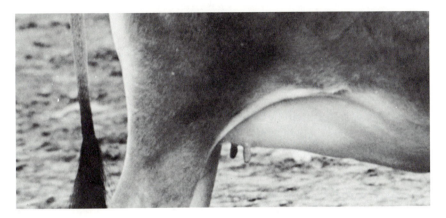

Figure 366 Excellent heifer udder with level floor; long, strong attachments; and teats of proper size and placing.

Figure 367 Almost perfect udder on a heifer, with long, strongly attached fore udder; level floor; and widely spaced teats of proper size and proper placing.

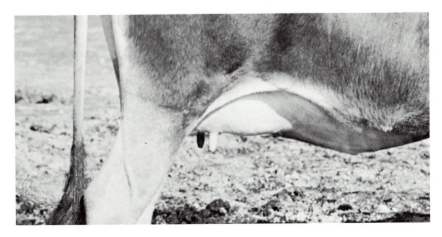

The above suggests that marked defects should be recognized and discriminated against in heifer judging. It is also important to be more critical on points that have little chance to improve with age and to be more tolerant of those that will improve as the heifer develops. Only a good judge with considerable practical experience can determine accurately the points of conformation that improve with age and, conversely, those that do not. Since overcondition exerts a harmful influence throughout the lifetime of the animal by decreasing production and affecting udder attachment, quality, and general udder shape, discrimination against it is a good example of practical application in judging.

The photographs presented in this chapter supplement the discussion and illustrate good type in young dairy animals. Also, as an example of proper explanation on placings, a set of reasons has been developed for a Holstein heifer class.

Reasons for Placings of Holstein Heifer Class
George W. Trimberger

This class of Holstein heifers is difficult to place; for the final decision they were lined up in the order of 1–4–2–3. This places the smoothest heifer with the greatest depth of body and most style, symmetry, and balance at the top of the class, and the coarsest heifer, lacking in clean-cutness, refinement, and dairy quality in fourth place, with a close middle pair of similar conformation.

One is definitely the best in the class and places over heifer 4 because she is stronger with deeper ribs both fore and rear. She also excels in spring of rib. She is much more attractive about the head, where she excels in balance and character, especially about the nose and forehead. Four is a trifle plain of nose and short of forehead. One also has a decided advantage in depth from withers to point of shoulder and in smoothness

Figure 368 Holstein heifer calf class.

1

Figure 368 (continued)

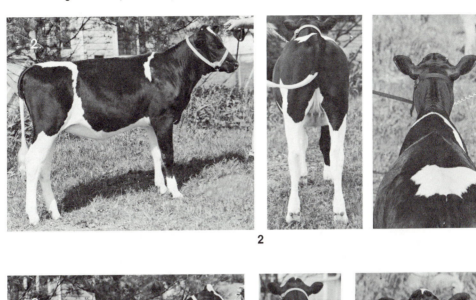

2

3

4

about the point; 4 must be criticized for being too prominent at the point and lacking in strength of shoulder attachment. Both 1 and 4 are excellent through the topline, but 4 must be granted an advantage in flatness and fullness of rump, especially when viewed from the rear. From this view 4 must also be granted a slight advantage in width of rear udder and sharpness at the withers; 1 is a trifle open here. In feet and legs 1 has a definite advantage, with stronger leg bones and pasterns. In comparison, 4 is long and a trifle weak in the region of the pasterns.

Four places over 2 for the closest placing in the class. The placing is difficult because 2 follows the general pattern of 1, but 4 is larger, with considerably more development at this age. She is deeper at the heart, sharper and smoother at the withers, where 2 is open and rough and the poorest in the entire class. Four has an undisputed advantage in fullness of rump and refinement of tail head; also, this heifer is the best in the class through the pelvic region. Both 4 and 2 have about the same leg conformation, but 4 has the advantage with a stronger leg bone, even though she had to concede this point when compared with the top heifer. It should be mentioned also that 2 has an advantage through the front quarters, since she displays more style, balance, and Holstein breed character about the head, particularly about the nose, angle of the jaw, and forehead, as well as depth and smoothness of shoulder, with more refinement at the point and a firmer attachment of shoulder.

Two places third and over 3 because she is much smoother throughout, with a considerable advantage in blending of parts, especially through the front part of the body. The style, symmetry, and balance of her head, her clean-cut and more refined neck, and the great advantage in smoothness of shoulder, particularly at the point, give her considerable advantage over 3. Three is plain in this region, with some coarseness about the head, which shows a receding forehead, a heavy, throaty condition underneath, and is attached to the body with a short, heavy, and compact neck lacking in dairy refinement. It is also immediately apparent that these two animals differ considerably in dairy refinement and general body type, and that 2 rather than 3 follows the type of the pair selected to head this class. There is a great difference in favor of 2 for smoothness at the point of shoulder, but 3 has an advantage at the withers, where 2 is again criticized for showing open and rough. Two is stronger and has an advantage in fullness of heart behind the forearm, and 2 is proportionately deeper in fore and rear ribs, where 3 is somewhat rounding. Both have good rumps, but 2 has a slight advantage in strength of loin.

In making this placing of 2 over 3 for the bottom pair, it is readily admitted that 3 is the largest heifer in the class, and 2 must grant her an advantage in strength of frame; this is indicated by strength of bone in her hind legs, where she has a slightly better set to the hock from both the side and back views. It was previously mentioned that 3 is also granted an advantage at the withers. In placing 3 last, it is recognized that she is the heifer with the most scale in the class and has a good rugged frame, but she is of a different type from the other three and lacks in dairy quality throughout, especially her coarse, plain, head; throaty, short neck; and heavy point of shoulder. She dips slightly at the loin and lacks in natural depth of rib, but must be given credit for the best set of legs and the sharpest withers of any heifer in the class.

Reasons for Placings of Holstein Heifer Class
William M. Etgen

I placed this class of Holstein heifers 1-4-2-3. One, a large, smooth-blending, well-balanced heifer, heads the class. In a close placing between two sharp, angular, but smaller heifers, 4 places over 2. Three, a large but coarser and thicker heifer, is in fourth place.

I placed 1 over 4 because of her advantage in overall size, scale, and substance. She is a larger heifer that stands taller at the withers. One is much deeper in the chest and fore and rear rib and has more spring of rib. She has more substance of bone, especially in her rear legs. Four is too refined in bone and 1 is stronger in her pasterns. One also has more breed character and style about the head, is tighter through her shoulders, and blends more smoothly at the point of shoulder.

In a closer placing, 4 places over 2 because she is more pleasing in general appearance and is a deeper, longer calf. She is a stronger-jawed heifer that is smoother through the shoulders, fuller in the crops, and more nearly level from hooks to pins. Four is deeper in her heart and has a longer body.

In the bottom pair, 2 places quite easily over 3 because of her tremendous advantage in dairyness. She shows more refinement about the head, is longer and much cleaner through the neck, is cleaner over the topline, and is more prominent in the hips and pins. Two has more style and femininity about the head and is flatter over the rump, especially when viewed from the rear. Three must be penalized for being low at the pins and thurls.

I grant that 3 is a larger heifer and has a considerable advantage in depth and length of body and is fuller in the crops, but I placed her last because of her lack of clean-cutness and angularity throughout.

CONCLUDING REMARKS

In the final analysis, the judging of heifers should be geared to a pattern that results in the preferred type of the mature cow. This suggests a heifer of adequate size, upstanding in general appearance, an open and deep-ribbed body conformation, with dairy quality and strength throughout; the heifer should be free from any indication of undesirable udder conformation and should be standing on a sound and correct set of legs and feet for good mobility during an anticipated long, useful life.

The pattern described above is shown in Figures 369 through 372, which are pictures of an individual bred and raised by a farmer-breeder of outstanding Holsteins in New York State.

Figure 369 A promising calf that has extreme dairy quality and superb strength without any sign of weakness and that has an upstanding, open-bodied conformation. Shown three times, she was twice Junior Champion. (Courtesy Collins-Crest Farm, Perry, NY)

Figure 370 Undefeated Junior Champion while developing as a yearling heifer, All-American Senior Yearling. Note combination of strength with quality. (Courtesy Collins-Crest Farm, Perry, NY)

Figure 371 This cow was undefeated in her class as a 2-year-old and was Reserve All-American. Her outstanding udder (added to a magnificent body) was Best Udder wherever shown. She produced 22,909 lb of milk and 781 lb of fat at 2 years of age and classified Very Good—89 First Classification. (Courtesy Collins-Crest Farm, Perry, NY)

Figure 372 This cow in mature form (Ex 96 2E) produced 31,028 lb of milk and 1020 lb of fat at 5 years of age. She was Reserve All-American 4-year-old the previous year. She has a production pedigree and produced over 20,000 lb of milk in each of her first four lactations. Although she is 60 inches tall, she is a refined kind of cow with excellent classification in every breakdown. (Courtesy Collins-Crest Farm, Perry, NY)

19

JUDGING GROUP CLASSES

Every animal in a group class should be outstanding in type if the group is to make a strong showing. Uniformity of proper type, with each animal cut according to the same pattern, carries considerable weight. It is especially important that individuals in the same group have no specific weakness in common, particularly in the get-of-sire classes. Maturity is a distinct advantage, provided that the individuals in the group carry their years lightly, indicate good wearing qualities, and that all animals in the group show proper development for their stage of maturity.

Group classes may present various problems for the judge. A group comprising uniform individuals of better-than-average type is usually given preference over one in which half the individuals are very outstanding and others are below the type standard. Normally, age and maturity have an advantage. However, if a group of animals is mature but is inferior in conformation and breed characteristics, the judge is well justified in rejecting these animals in favor of others that are outstanding in type, even though somewhat lacking in maturity. Disqualified animals are not eligible to be shown in group classes.

A junior get-of-sire class with one or two bulls is usually less appreciated than a class of four heifers. For this class, yearling heifers have a distinct advantage over calves; if the yearlings are inferior, however, the judge is justified in selecting a younger group provided they are uniformly top individuals.

Dry cows in a get-of-sire group should not pose too much of a problem if they are evaluated in accordance with the recommendations given in Chapter 16. But a group of four cows for the dairy herd should comprise four cows in milk, in spite of the fact that 16 percent of the cows in a normal herd are usually dry. Four good aged cows have an advantage over younger cows that have not been properly tried for wearing qualities. A dairy herd with one or more dry cows is at a disadvantage even if the dry cows are superior in conformation, and the rules at many shows distinctly specify that the four individuals in the group must be in milk. In more recent years many dairy shows have discontinued some or all of the group classes in favor of a Breeder's Herd class or Breeder's Herd of Five Females class. In this class individual excellence is of paramount importance.

For close placings in the group classes, the animals in each group should be lined up in tandem arrangement (head to tail in a straight line) to allow for better detailed comparison. Individual class placings should be kept in mind

from the previous classes rather than obtained from the clerk. When finally placed, each group should be lined up abreast, one group behind the other in order of placing.

Figures 373 through 380 illustrate almost perfect uniformity of group type and set a pattern for recognition of outstanding groups.

Figure 373 Outstanding udders, dairy character, and uniform correctness of type typify this nominee for All-American Get-of-Sire. The sire was a popular high-predicted-difference sire. Three of the four composed the All-American best three females (Fig. 376). (Courtesy Charles E. Cotting, Berlin, MA)

Figure 374 The daughters of this Holstein bull were voted All-Time All-American Get of Sire after his daughters won the All-American award for five successive years. The group was selected on the basis of individual superiority, uniformity as a group, and maturity. There is not a weak spot on any individual or on a group basis. Each individual's merits are listed from left: EX 92, All-American 2-year-old, Reserve 3-year-old, and High Honorable Mention All-American 4-year-old; over 18,000 lb milk at 2 years, 7 months. Next: 2E 95 All-American 4-year-old and Reserve All-American Aged Cow, 23,591 lb milk at 5 years, 11 months. Third: 2EX 97 Reserve All-American Aged Cow, 27,630 lb milk at 6 years, 6 months. Right: 5EX 97, voted four times Reserve All-American and two times All-American Aged Cow and High Honorable Mention All-Time All-American Aged Cow. She has four records of over 30,000 lb of milk. (Courtesy A. B. Baker, Canton OH; Hanover Hill Holsteins, Armenia, NY; Pinehurst Farms, Sheboygan Falls, WI; and Allen & Roy Hetts, Fort Atkinson, WI)

Figure 375 Uniform high quality and correctness characterize this undefeated All-American Junior Get of Sire. (Courtesy C. Robert Sibley and family, Mt. Carrol, IL)

Figure 376 These All-American Best-Three Females are also included in the Get of Sire in Fig. 373 that was nominated for All-American. (Courtesy Charles E. Cotting, Berlin, MA)

Figure 377 This outstanding All-American Best-Three Females contains exceptional individuals of uniform high quality; an All-American 4-year-old, a reserve All-American 3-year-old, and a 3-year-old that placed first at two major shows. (Courtesy Pinehurst Farms, Sheboygan Falls, WI)

Figure 378 This produce of dam won All-American honors for three consecutive years and later was voted All-Time All-American Produce of Dam. The first cow (Roxy) is 4E 97 GMD with four records over 1000 lb fat. She has 15 EX offspring and was a member of an All-American Get of Sire. Both cows were EEEE across on udder. Rocket, the second cow from the camera, is 3E 96 with 21,460 lb milk, 3.9%, 835 lb fat at 5 years, 2 months on 2X in 365 days. (Courtesy Robert and Craig Miller Dundee, IL, and Lorne Loveridge and family, Grenfell, Saskatchewan, Canada)

Figure 379 Progeny representing a potent transmitter for production and type. These four paternal sisters represent a sire with a high predicted difference that is one of the best for the breed in milk production. His daughters are outstanding in type. The group in this picture were designated All-American after they were winners at the National Show. They represent the product of a very successful breeding program. (Courtesy American Jersey Cattle Club, Columbus, OH)

Figure 380 This All-American breeder's herd contained four individuals that were All-American nominees, including the All-American and Reserve All-American aged cows (the cows on the upper left and right, respectively). The cow on the center row left was sixth place 2-year-old at the National Show. The heifer in the center row right was Reserve All-American Junior Yearling and the one on the lower row was High Honorable Mention All-American Junior Yearling. They are fine examples of uniform quality, the kind it takes to make a winner. (Courtesy Palmyra Farm, Hagerstown, MD)

20

SHOW-RING TECHNIQUES
AND PROCEDURES

It is important for the professional judge to supplement his or her knowledge of cattle and the preferred type with accepted and standardized show-ring techniques and procedures if the judge wishes to be both popular and successful. In many situations the judge, with the cooperation of the person at the halter, is the whole show. A really capable superintendent of cattle, however, quietly supervises all details and can make the show a smooth, interesting, and educational affair, to the general satisfaction of everyone present.

The working procedure and the attitude of the judge should be such that the judge develops confidence among the exhibitors, and the cattle superintendent should do as much of the policing as possible. The judge is the leader on judging day and frequently sets the pattern for the attitudes of the participants.

A judge must be completely honest and should also develop the reputation of treating both the exhibitors and the cattle as though he or she had never seen them before. Only in this way can a judge make an efficient, fair, and unbiased evaluation of the cattle.

The American Guernsey Cattle Club has published a booklet, *Showing and Judging Procedures*. This booklet has been very influential in standardizing judging procedures and in eliminating many of the problems that have been so prevalent in the show ring. As a result, both the American Dairy Science

Association and the Purebred Dairy Cattle Association have adopted these recommendations of procedures.[1]

BEFORE THE SHOW STARTS

The judge should be at the show at least 30 minutes before the scheduled starting time. This will give the judge time to confer with the cattle superintendent and/or the managers of the show. The judge can then check on the maximum numbers to place in the various classes, determine the facilities for giving oral reasons, and decide where the classes should be lined up so that the animals will be displayed to best advantage for the ringside. It is an advantage for the judge to be familiar with the show classification and the total number of classes, as well as the approximate number in each class. It is always appropriate for the judge to check with the management on the time schedule for the ring, for breaks at mealtime, and for closing time.

A judge should never go into the show barns before judging and should not fraternize in any way with the exhibitors, either at the show or away from it.

SHOW-RING PROCEDURE

The judge should require a high standard of conduct on the part of the leaders while in the ring. The judge should make requests or suggestions courteously and should not resort to disciplinary measures except in the rare instances when they are justified and necessary.

The judge should make a special effort to start the show at the scheduled time and keep it running smoothly by timing the classes properly and by making prompt, thorough decisions. Spectators and exhibitors lose confidence in a slow judge and the whole show drags. A slow judge soon fails to receive requests for his or her services. Nevertheless, a judge must take enough time to evaluate each animal accurately. A judge who works too rapidly often overlooks superior individuals and is likely to make serious placing errors. To save time, many judges begin to study a new class as it enters the ring. This also encourages exhibitors to bring their animals to the ring promptly.

[1]The cooperation of the American Guernsey Cattle Club in granting permission to use some of the information in this booklet is appreciated.

Entries should move clockwise around the ring, with the judge directing them by hand motions. Some judges prefer to start by thoroughly studying the class from a distance. Another method is for the judge to observe each animal by standing directly in front of it, then stepping to one side as the animal moves toward him or her, and then moving behind the animal and over to the other side as the animal slowly moves by. When each animal has been observed and analyzed in this manner, the judge moves quickly around the ring to recall certain features and to fix the overall picture in mind. The judge usually then motions for the class to come to rest so that each animal can be observed in the same way while in the standing position.

METHOD OF INSPECTION

From the front view, the shape and character of the head and neck, width and depth of chest, and set of legs and feet are observed. The side view permits observations on general style and symmetry; profile of the head; length and leanness of neck; depth of fore and rear ribs; length of barrel; chest capacity; shoulder conformation; topline, including length and levelness of loin and rump; udder size, shape, and texture; set of legs; and many other points of conformation. Observations made from the back view include sharpness of withers; smoothness of shoulders and point of shoulder; spring of fore and rear ribs; width and levelness of loin; flatness and width of rump, especially width of hooks, thurls, and pins; and set of legs, particularly width at hocks and leanness of thighs.

Finally the animal is viewed from the opposite side. Natural defects must be observed and remembered. The entire procedure must be handled with dispatch because for large classes at large shows only about one minute or less can be spent on each animal. This means about 20 seconds or less for observing each animal while in motion and another 20 seconds or less for observation while the animal is standing. Such rapid judging requires eyes carefully trained to see many things at a glance, as well as the ability to make, and retain, a mental record of each observation.

FORMING THE LINEUP

The class should be lined up in order in the center of the ring. For large and difficult classes, or if insufficient space is available, it may be an advantage to assemble a group of individual contenders side by side before the final ranking is made. This should be done in a different location, and preferably at a different angle, from that used for final placing. In extremely large classes this procedure may be repeated two or three times. In large classes it often is an advantage to do this sorting while the class is in motion, but if space is limited it can be done at any time. This limitation of space may prompt a judge to dismiss part of the class back to the barn. Such action is discouraging to an exhibitor, however, and to avoid placing a stigma on these animals, it is usually better to allow them to remain at one side of the ring. Thus, the exhibitor may stay and watch the proceedings if he or she wishes, and the animals can still be considered a part of the class. It also gives the judge an opportunity to reinspect the group; perhaps the judge may even use one or two of these animals in the final lineup.

The first-place animal of every class should always stand in the same ring location. After placing the animals and lining them side by side, it is a good practice to have close placings led out of the lineup so that they can be compared again before making the final decisions. This technique conveys to the ringside some of the problems in making placings. While the closely placed animals are being led out of the lineup, or at other convenient times, a special examination can be made of each strong or weak point of conformation to indicate to the exhibitors and the ringside the differences that exist in certain individuals. The two individuals to be compared can be led out either in front or behind the class, depending upon available space.

Every possible effort should be made to avoid many changes in placings after the entries have been lined up, but occasionally it is necessary, and it is better to make a change than to make an obvious mistake. In changing the order, a tactful judge will move one animal up in a class rather than move another one down. Only on rare occasions should a switch be made in the top pair. The judge should feel confident of the top placings before designating the order; occasionally, however, the animal selected for first place "falls apart" after standing at rest for a while. To check on this, it is best to walk in front of and behind the entire class before the final decision is indicated. Some animals fail to retain their poise and form and may sag in the topline, open up in the shoulders, drop out at the point of the shoulders, or "fall apart" in several of these or other parts of conformation. A judge cannot be satisfied with the final lineup if such a situation occurs. The judge has previously noted the animal's conformation while it was moving and has evaluated it with a composite appraisal to allow for defects in a standing position. If the animal later shows serious defects after standing, a change is justified, and it will be favorably received by the exhibitors and the ringside. To avoid such a situation, though, the top-placing individual should be studied and observed thoroughly before starting the lineup, particularly at a large show where competition is tense and the top placing carries so much importance. A successful and experienced judge will not try to appease anyone who complains in the show ring.

It is very important to have all the individuals in the class walk around the ring after they have been lined up in the center. Because of this maneuver, a lineup from right to left (Fig. 381) is normally preferred. It is very inconvenient to move an entire class from a left-to-right lineup (Fig. 382).

When the judge is satisfied with the order of placing, the judge indicates by a sweep of the hand to the clerk, exhibitors, and ringside that a final decision has been made.

Figure 381 Class of outstanding senior yearling Jersey heifers lined up from right to left. Note how easily the class can be moved clockwise around the ring from this position, in contrast to the lineup shown in Fig. 382. (Courtesy American Jersey Cattle Club, Columbus, OH)

Figure 382 Brown Swiss cow class lined up from left to right. This lineup can be justified if it is more convenient for the space available for judging. Usually the right-to-left order is preferred for ease in moving the class. (Courtesy Brown Swiss Breeders Association, Beloit, WI)

REASONS FOR PLACINGS

It is one of the responsibilities of a judge to give oral reasons; no dairy cattle show should be held without arrangements for the judge to justify the placings. Loudspeaker facilities should always be available for this. To give effective reasons, a judge should be trained to remember the points of conformation he or she has observed. The judge should develop the proper vocabulary to make a good impression with positive and convincing reasons. A good judge will explain graphically to the exhibitors and the ringside the particular points on which one animal has an advantage over the next. This can, and should, be done without criticizing, which is annoying to an exhibitor, especially if his or her animal is a previous winner. It is much more tactful to praise one particular animal or a part of conformation than to criticize the bad points of another animal.

Reasons for placings should be short and concise and they should stress only the critical differences of conformation. Thus the time spent in giving them will take very little of the total time allotted for the show. They can be given while the clerks pass out the ribbons and record the placings and while the cattle are moving out of the ring—still another way in which a judge can demonstrate leadership in working with responsive exhibitors to the general satisfaction of the ringside.

SUMMARY

Show-ring techniques and procedures for judges are summarized below:

1. Arrive at the show at least 30 minutes before the scheduled starting time.
2. Confer with the management of the show on classification, number to be placed, equipment for giving oral reasons, and best place in the ring for the lineup.
3. Do not visit the barns or fraternize with exhibitors.
4. Maintain ring discipline but do not embarrass anyone.
5. Move the class clockwise around the ring.
6. Proceed in an orderly way to analyze each animal, noting its good points and its defects.
7. Make a careful observation from the front view, side view, and back view as the animals are slowly led around the ring; repeat this procedure while they are standing.
8. Make prompt and thorough decisions.
9. Line up the animals from right to left in the center of the ring, with the front feet on a higher spot and with the side, quarter side, or rump toward the ringside.
10. If all or part of the class is moved in order around the ring, bring it back to the same place.
11. Lead out and inspect very close placings.
12. Walk along the front and return behind the lineup.
13. Make any changes that are necessary, but keep these to a minimum.
14. Motion to the clerk that the class is finished.
15. Give oral reasons over the loudspeaker.
16. Lead champion animals in order of age.

21
FITTING AND SHOWING DAIRY CATTLE

Owners of purebred dairy cattle, including many young people, frequently participate in dairy cattle shows and sales, both of which are excellent means of advertising and promoting purebred cattle, contacting prospective buyers, and socializing with other breeders. Sales can also be a means of learning to estimate the value of animals more accurately. In addition, shows offer the opportunity to learn more about the desired type of dairy animals and how to prepare and show them to best advantage. Showing gives many dairy farmers and young people the opportunity to participate in a competitive event while making new friends among fellow breeders and learning more about type. Shows are responsible for increasing many young people's interest in dairying.

Since one of the major purposes of exhibiting dairy cattle is advertising and promotion, it is important to show your best animals. The two essentials of exhibiting are that the animals be (1) competitive for the level of the show and (2) properly fitted and shown to best advantage.

The exhibitor should prepare and exhibit only those animals that have desirable dairy type, that is, good enough to be competitive for the show in which the exhibitor plans to participate. There is considerable difference in the degree of excellence of a dairy animal that is required to exhibit at a county fair or at a national breed show. To select those individuals, one must have good knowledge of dairy type and must be familiar with the level of competition at the show in which he or she plans to participate and in the show classes for that particular show.

If one is to prepare and present dairy animals to their best advantage, one must learn how to fit and show them. This includes feeding, washing, clipping, training to lead, and many other details, all of which we will discuss in this chapter.

SELECTING ANIMALS TO SHOW

Selecting animals to show involves knowing the show age classes, selecting the animal(s) that are best for that class, and selecting only those that are competitive for the show in which they are to be entered.

Shows vary somewhat in ages and other qualifications for the various classes, but they usually follow rather closely the Purebred Dairy Cattle Association's Recommended Show Classifications. Not all dairy shows adhere strictly to these classes and ages. Some shows have additional classes, some have fewer classes, and some have different requirements for certain classes, for example, some require only three animals in get-of-sire classes. Therefore, it is always wise to check the catalog for each show very carefully.

Older animals generally have an advantage in calf and heifer classes because older animals are larger. In the milking-age class, fresher cows that show more bloom of udder and more dairyness usually have an advantage over stale, thicker cows. These factors should be kept in mind when selecting animals to show. This does not mean that one should show animals that have inferior conformation just because they are older or more recently fresh. Younger calves that have correct type or stale cows that have good conformation and a fine udder are superior to older or fresher animals whose conformation is faulty. Good udders are of particular importance in selecting cows to show. Straight toplines, good legs, and dairyness, along with good size for age, are important when selecting heifers.

It is not necessary to have an entry in every class. Some dairy farmers show inferior animals just to have an entry in every class or to have sufficient animals for all the group classes. This practice should be discouraged because it is contrary to the major purpose of showing. Displays containing inferior animals are not good advertising.

It is often advisable to request an opinion from a respected breeder concerning which animals to show because people often tend to be less than completely objective about their own animals. This is especially true of inexperienced exhibitors. Early selection is also important if one wants to have adequate time to fit and prepare the animals. This is especially true for underconditioned or overconditioned animals.

FITTING DAIRY CATTLE

The major purpose of fitting dairy animals is to have them looking their best at the time they are shown. To achieve this purpose, (1) evaluate the animal carefully to identify both weaknesses and major strengths, (2) devise a plan to correct weak points so far as possible and emphasize strong points, and (3) implement this plan. The last two points generally involve planning a feeding program, training to lead, foot care, washing, and clipping.

Evaluation

The purpose of evaluation is to identify problems that can be corrected completely or to a large extent before show time. To do this properly, select the animal several months before the first show. In the case of milking animals, planning may start a year or more in advance in order to have the animal freshen at the correct time. Evaluation of an animal consists of carefully examining the animal to determine what can be done to improve her weakest areas. As an example, often heifers are overconditioned or underconditioned. Correcting this requires changing the feeding program. Depending on the degree of overconditioning or underconditioning, it may take several months to correct the problem.

Overconditioned heifers should usually be placed on a diet of high-fiber forages and minimal, if any, concentrate feeds. Care must be taken to ensure adequate intake of protein, minerals, and vitamins lest nutritional deficiencies or restricted growth occur. Thin heifers should be fed a diet relatively high in low-fiber, high-energy feeds. This diet generally includes more high-energy concentrates and low-fiber forages. Heifers that are adequate in size but lacking in depth of body or spring of rib can often be improved by feeding them a bulky ration such as a combination of hay and beet pulp.

Animals may have feet or leg problems that can be alleviated to varying degrees by proper hoof trimming. It is best to trim the hooves long before show time because many animals may need corrective trimming, which may take many sessions over several months. Also, since animals sometimes become temporarily lame after trimming, it is advisable to identify and correct foot problems early.

It is also recommended that animals be thoroughly washed and examined for symptoms of diseases and for external parasites such as ringworm, warts, lice, and mange several weeks or months before show time. If discovered early, these skin conditions can usually be healed well before show day. An increasing number of shows are prohibiting animals from being shown if they exhibit symptoms of one or more of these skin diseases.

It is wise to check the health regulations of the shows one plans to attend at the time of evaluation of the animals. Most shows have very specific requirements on certification of freedom from infectious diseases such as tuberculosis or brucellosis.

Feeding

In addition to feeding to correct underconditioning and overconditioning, feeding from birth, just prior to moving to the show, and while at the show can contribute significantly to making the animal look her best on the show day. It is difficult to fit an animal properly unless she has been properly fed since birth. It is important to feed a ration that encourages maximum growth without overconditioning. Stunted calves rarely reach adequate size and are heavily penalized in the show ring. Conversely, calves that are overfed and very fat from birth can rarely be trimmed down to an ideal stage of condition and they often remain throaty, thick through the thighs, and patchy over the rump and pins. There are many good booklets explaining how to feed calves properly for growth without overconditioning.

We recommend that dairy animals that are to be shown become accustomed to the feeds and feeding methods that will be used at the show before the animals are moved to the show. If hay and beet pulp are to be fed at the show, the animals should gradually become accustomed to those feeds before leaving home. If they are to be fed and watered from buckets and tubs at the show, it is advisable to feed and water them from buckets and tubs for several days before moving the cattle from the farm. Moving cattle from their home surroundings is an upsetting experience for many of them. This can often be greatly reduced and many can be prevented from going "off feed" by adjusting them to similar conditions before they leave home. Tying and feeding them the feeds they will receive at the show from containers that will be used at the show will help prevent many problems.

Training

If an animal is to look its best, it must be trained to walk slowly, take short steps, keep its head up, stay alert, and respond to a light touch from the leader. The animal should be trained to stand with its weight evenly distributed on all four feet and with its feet in the proper position. Proper leading and posing can alleviate many minor conformation problems, for example, crooked legs, loose shoulders, and weaknesses in the crops or loin. These minor problems can also be accentuated with improper handling.

Training an animal to be led and to pose properly should take place well in advance of the show season. Since it is generally easier to teach younger animals to be led and to pose properly, many purebred breeders routinely train all their heifers to handle with a halter at a young age.

Training usually begins by tying the animal with a rope halter to a wall

Figure 383 When showing a cow, pose her so that the rear leg nearest the judge is slightly forward. This covers any quartering on the floor of the udder. It also presses the udder away from the judge, thus making the udder appear tighter in its attachment. (Courtesy Dennis A. Hartman, Blacksburg, VA)

Figure 384 When showing a heifer, pose her so that the leg nearest the judge is slightly back. This makes the heifer appear longer bodied and more correct in set of hock. (Courtesy Dennis A. Hartman, Blacksburg, VA)

that has a smooth surface. Almost all animals soon become accustomed to the halter; the smooth surface prevents them from injuring themselves. The animal should be tied with a slipknot that can be loosened quickly if the animal should fall or throw itself while fighting the halter. It is also advisable to work with the animal during this initial tie-up period. Brushing the animal while talking to it gently usually helps quiet it and helps accustom it to being tied and handled. After a day or two of being tied, leading training can begin. This is best achieved by coaxing and gentle handling, not by dragging the animal behind a vehicle or using physical abuse. Leading the animal back and forth to water is often helpful. Gentle yet firm control of the animal during the training period is recommended. Daily training sessions of short duration in being led and in posing properly are usually more effective than fewer and longer sessions of pulling, pushing, and tugging. A leather halter like the one that will be used in showing should be used during the last part of the training period in order to accustom the animal to a leather halter.

An often overlooked part of the leading training program is leading each animal to help overcome its particular faults. It is generally recommended that an animal be led so that its head is held high, but this detracts from the appearance of an animal that has a weak loin. It is also usually recommended that a dairy animal be posed with its front feet even when viewed from the side, but if an animal is a bit loose at the point of shoulder, it is often beneficial to have the front leg nearest the judge slightly behind the other because this makes the shoulder appear less prominent. The point to be emphasized is that all fitting practices, including leading, should be designed to make each individual animal look its best. The important points to consider in proper leading in the ring are covered in Chapter 22.

Foot Care

The hooves of all dairy animals should be in proper condition for the health and appearance of the animal. Good foot care is essential if the animal is to walk and pose properly. Long toes and misshapen hooves make it difficult for an animal to walk or pose properly. Hooves should be trimmed well in advance of the time for showing in order to prevent possible lameness or foot tenderness on show day. It is becoming increasingly popular to hire a professional to trim the feet because it is a very difficult job, especially on adult animals, without the proper equipment, especially some type of chute or trimming table where the animal can be harmlessly restrained during the trimming process. It is much easier to trim heifers' hooves. It is important to note that trimming is much easier if the animal has been trained to be led and to pose properly, if the hooves are relatively soft (animals having been on spring pasture or animal's feet having been placed in a foot bath for an hour or so before trimming), and if one has a sharp wood chisel or two, a rubber or wooden mallet, a sharp hoof knife, and a rasp. Figures 385 through 388 explain the process.

Figure 385   A chisel and rubber mallet are used to shorten the toes. It is important to have a piece of plywood under the hoof to avoid damage to the chisel. (Courtesy Dennis A. Hartman, Blacksburg, VA)

Figure 386   A chisel with the handle cut off and a bar welded across it allows one to grip the chisel securely and trim the sole of the hoof. The workbench is useful for holding the animal's knee. The same technique can be used to trim the front and rear feet. (Courtesy Dennis A. Hartman, Blacksburg, VA)

Figure 387   The rear feet are more difficult to hold. The technique shown is one of the best. Some prefer to use a bench or bale of straw on which to rest the foot to gain additional stability. (Courtesy Dennis A. Hartman, Blacksburg, VA)

Figure 388   A rasp is also useful to smooth and shape the hoof. (Courtesy Dennis A. Hartman, Blacksburg, VA)

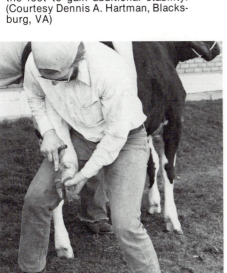

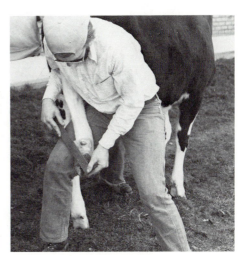

Figures 389A and 389B illustrate what can be achieved through proper fitting and training. Figure 389A is a photo of a heifer taken the day after she was brought in from pasture. Figure 389B shows the same heifer just after she placed third in a large class at a State Black and White show.

A

Figure 389 (A) This tall, long, basically correct heifer was chosen to be fitted and shown as a junior yearling. Careful evaluation indicated a weak loin, high tail head, throatiness, a shallow body, and poor response to handling, as indicated by her pose. (B) The finished product after washing, training, clipping, and so forth. The difference in general appearance and style is remarkable. She also appears cleaner in her throat and deeper and wider in her rib. (Courtesy VPI & SU Dairy Center, Blacksburg, VA)

B

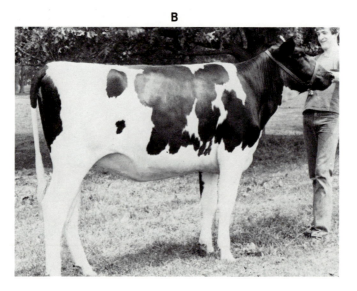

Washing

Show or sale animals should be thoroughly washed at least 4 to 6 weeks before showing to help detect any skin problems, and every week or two thereafter to keep them reasonably clean. They should be washed soon after arriving at

A

Figure 390   This animal is being washed thoroughly with a mild soap (A) and then rinsed thoroughly (B). Many professionals use a power sprayer because it speeds up washing and makes it easier to do a good job. Note that the animal is tied with a chain, not with a rope halter (B). (Courtesy VPI & SU Dairy Center, Blacksburg, VA)

B

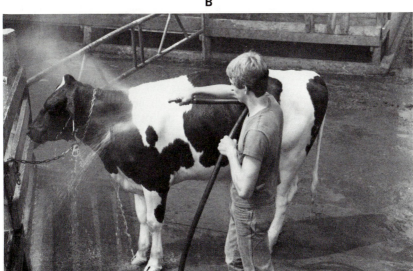

the show, as often as needed to keep them clean at the show, and again several hours before the first class. Most experienced exhibitors prefer to wash them within 4 to 12 hours before the first class.

The purpose of washing is to clean the animal, remove manure and other stains, and obtain a shining hair coat. When washing an animal, it is best to fasten a chain around the animal's neck and tie the animal to the wash rack. A chain that has a snap at one end enables one to use the same chain to wash animals of varying ages and sizes. A rope halter should not be used because it will shrink when it gets wet. This can injure the animal and make it difficult to remove the halter. The animal should be soaked with water, thoroughly scrubbed with a mild soap, such as castile, or with a soap especially designed for washing animals, and then rinsed until all of the soap is removed from the hair. Power sprayers greatly speed up washing and do a more thorough job, especially of rinsing all of the soap from the animal.

Some professionals also use a hot-air dryer to blow dry the hair, especially over the topline, because blow drying makes the hair stand up. They later trim this hair to help smooth out uneven places over the topline and clip the withers to a sharp point and elsewhere to improve other weaknesses in conformation.

It is advisable to keep the animal blanketed with a light flannel blanket and to avoid exposing it to cold drafts until it is dry. If an animal has very long hair, it might be advisable to keep it heavily blanketed for several weeks before the show so that the unsightly long hair will loosen and can be removed by brushing. After the last washing before showing, the blanket should be removed an hour or two before entering the ring. This allows time for the last-minute touch-up, clipping over the topline and brushing to bring out hair-coat sheen.

Clipping

The purpose of clipping is to improve the animal's appearance. When properly done, clipping enhances strong points and helps disguise weak points. It is one of the most important parts of fitting an animal for sale or show. A good job of clipping can make the topline appear straighter and sharper, the rump more level and longer, the neck longer and cleaner, the head cleaner and more attractive, and the legs cleaner and straighter, and it enhances the appearance of the udder. These changes can make a big difference in the appearance of the animal and ultimately in the placing in the show ring or the selling price at a sale.

As in many other fitting practices, clipping should be tailored to the individual animal. Areas normally clipped include the head, ears, neck, and for cows, the udder. Clipping parts of the topline, shoulders, and legs usually enhances certain parts of the animal's appearance. The initial clipping, especially on heifers, should take place soon after selecting them for showing but after they have been trained to be led and to pose. This will give the exhibitor the opportunity to evaluate the animal's show potential more accurately.

Each animal should be washed before clipping, because washing removes dirt that dulls clipper blades. It also improves the probability of doing a good job of clipping because the clean hair lies more smoothly and can be brushed up as needed to properly clip the topline, the rump, and the top of the neck.

The next step is to analyze carefully the animal to be clipped. The exhibitor must decide how to clip each animal to best enhance her appearance, where to clip, how much hair to remove at each place, where to leave some hair, and how to blend the clipped and unclipped areas so that no distinct lines are visible.

Many experienced exhibitors prefer to start at the tail because this part is relatively easy to clip and is of minor importance. In addition, clipping the tail first often accustoms the animal to the clippers so that she is less nervous when other parts of the body are being clipped. The tail should be clipped against the grain of the hair from about 3 to 4 inches above the beginning of the switch to the tail head. The clipped and unclipped areas near the tail head should be blended. The sides of the tailhead should be clipped, but the hair on the top of the tail head should be left unless the tail head is very high. This and other steps in clipping are illustrated in the Figures 391 through 399.

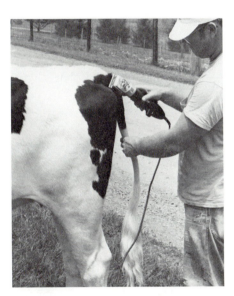

Figure 391 Clip the tail and tail head. Start at a point from 4 to 6 inches below the left hand of the person doing the clipping and clip against the grain to the tail head where the clippers are located. At this point, blend the long and short hairs. (Courtesy VPI & SU Dairy Center, Blacksburg, VA)

Figure 392 Turning the clippers over is the correct way to blend the clipped and unclipped portions of the rump. This technique also makes the rump and tail head appear more level, for it removes the hair from the high spots. It is wise to brush the hair against the grain in the low spots (hair spray is often used) before trimming. A similar procedure is used to improve the appearance of the topline and loin so that the result is as level a topline as possible. (Courtesy VPI & SU Dairy Center, Blacksburg, VA)

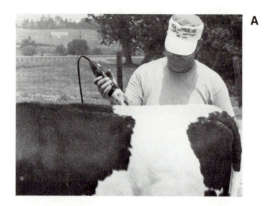

A

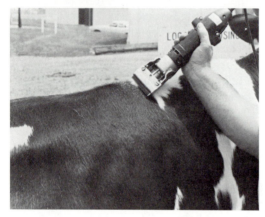

B

Figure 393 To clip the topline, brush the hair against the grain and then remove hair from the high spots by turning the clippers over and carefully trimming. (Courtesy VPI & SU Dairy Center, Blacksburg, VA)

Figure 394 Clipping the rear legs by clipping against the grain on both sides of the hock can make the animal appear to have a cleaner, flatter bone (A). Be sure to blend the hair with the unclipped area above and below the hock (B). If an animal tends to be sickle-hocked, clipping the back side of the hock can make this problem appear less severe. Do not clip too low on the leg, because it may accentuate the problem. (Courtesy VPI & SU Dairy Center, Blacksburg, VA)

A

B

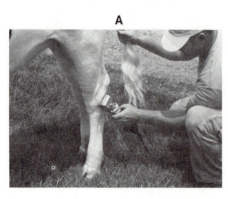

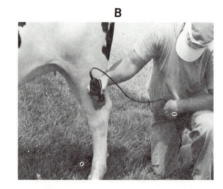

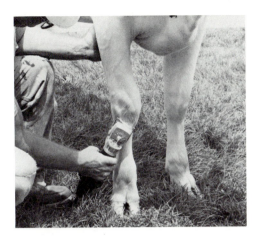

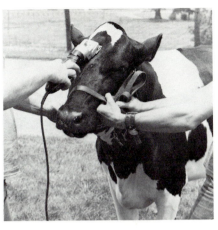

Figure 395 Clip any long or stained hair from the knees. Again, be sure to blend the clipped and unclipped areas. In this photo, the heifer's left knee has already been clipped. Note how much cleaner it is and how much cleaner the bone in her left leg appears. (Courtesy Dennis A. Hartman, Blacksburg, VA)

Figure 396 Clip the head by clipping against the grain. Clip as short as possible. Clip the inside and outside of the ears of all breeds except Brown Swiss; leave the hair inside their ears. Clipping the head greatly improves cleanness and breed character. (Courtesy Dennis A. Hartman, Blacksburg, VA)

Figure 397 Clip the neck back to the point where the clippers are positioned in (A). Do not remove the hair on top of the neck just in front of the withers. Most animals have a low spot in this area that can be corrected by properly shaping the hair just in front of the withers. Clipping the shoulders and withers is very critical. It is important to have the animal standing squarely on her front legs before starting. Start clipping at the point of shoulder as shown in (A). Clip at an angle toward a spot slightly behind the point of withers. As you approach the top of the shoulders and withers, gradually take the clippers away from the shoulder (B). Leave the hair at the top of the shoulders and the withers so that you can blend the shoulders smoothly into the withers and chine area and shape the hair over the top of the withers. (Courtesy VPI & SU Dairy Center, Blacksburg, VA)

A

B

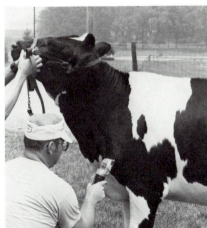

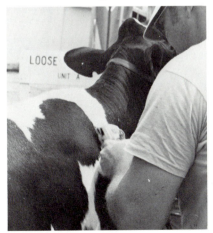

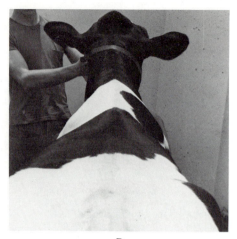

A   B

Figure 398 Blend the clipped and unclipped areas of the shoulders over the blades and at the top by turning the clippers over and clipping with the grain of the hair (A). Leave some hair on top of the withers, brush it against the grain, and then shape it to a point to accentuate the appearance of sharp withers, which indicates dairy character (B). Then shape the hair in the low spot in front of the withers by brushing it against the grain and shaping it to a point on top and level with the withers and the rest of the topline. (Courtesy VPI & SU Dairy Center, Blacksburg, VA)

Figure 399 The udder is clipped on both dry and milking cows because clipping enhances the appearance of udder quality and makes udder veining more prominent. Clip the entire udder. Clip the rear udder to a high point where it joins the body and blend smoothly. It is sometimes recommended to clip the hair on the inside of the thighs to give the appearance of a wider rear udder. The fore udder attachment is often difficult to clip, especially if it is faulty. If the fore udder attachment is smooth and strong, clip the fore udder closely and blend the attachment into the body wall. If it is a bit weak or bulgy, it is often wise to leave some longer hair in the area of the attachment and blend it smoothly between the fore udder and the body wall. The milk veins should be clipped to improve the milky appearance of the udder. There is some disagreement on whether or not the belly should be clipped in milking-age animals. This practice is much less popular than it was some years ago. (Courtesy Dennis A. Hartman, Blacksburg, VA)

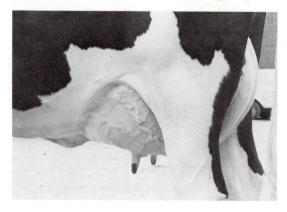

HANDLING DAIRY CATTLE AT THE SHOW OR SALE

Getting the cattle safely to the show or sale and presenting them to their best advantage are the final steps in fitting and showing dairy cattle. Below are recommended procedures.[1]

Hauling

The transporting of the animals to the show in such a manner that they arrive uninjured and healthy is an important part of successfully showing cattle. Overcrowding, slippery floors, faulty loading chutes, or poor loading conditions enhance the risk of bruising and injury, and must be avoided. Squeezing ten animals in a truck only large enough for eight is a poor practice. It is better to make two trips. Surfaces should be of the nonskid type and/or covered with sand, sawdust, or shavings. Inside surfaces should be free of projections and sharp edges. Loading chutes should be of strong construction, have a skidproof surface, and fit properly to the truck entrance. The truck should be in good condition and carefully operated. Sudden starts, stops, and speeding around curves must be avoided.

Many experienced dairy cattle exhibitors prefer to transport cattle when they are relatively empty and avoid feeding them large amounts of concentrates and succulent feeds within 12 hours of transporting. Cattle handled in this manner defecate less during hauling, and so are cleaner when they arrive. This practice may increase their appetite on arrival and ease problems with getting the cattle on feed in the new surroundings.

Arrangements should be made to arrive at the show a minimum of one to one and a half days, and preferably two to three days, before the show. This allows time for the animals to adjust to the new surroundings and get back on feed, and allows ample time to get the animals in the best shape before showing.

At the Show

The comfort of the animals should be the first concern after arriving at the show. They should be tied in a freshly bedded stall. Immediate needs such as feeding, watering, and milking, depending on the distance traveled, should be taken care of and a regular routine established for these necessities. Cattle usually prefer to lie down and perhaps eat a little hay after unloading. The animals should be washed as needed, or brushed to enhance their appearance.

Supplies and tack should be stored neatly and conveniently and the display area made and kept neat, orderly, and attractive. One of the pur-

---

[1]Reprinted by permission from *Dairy Cattle Feeding and Management,* 6th ed., by W. M. Etgen and P. M. Reaves, Copyright, John Wiley & Sons Inc., New York, NY, 1978.

poses of showing is advertising and promotion of cattle. Another especially important promotional consideration at fairs attended by urban people, our milk and dairy products customers, is promotion of milk and dairy products. Both purposes are greatly enhanced by the neat, attractive appearance of the total display; by clean, contented, high-quality animals; and by courteous personnel caring for the animals. Conversely, dirty cattle, littered areas, and discourteous attendants detract significantly from the purpose of showing and must be avoided.

Final Preparations

After the comfort of the animals has been cared for and the display area arranged, the entries should be rechecked to make sure everything is in proper order. The rules and methods of the particular show should be studied and every detail carefully watched. The showing schedule with approximate times should be learned, for this will determine when certain cattle-preparation tasks will need to be carried out. The purpose of all the prior preparation such as fitting, clipping, and so on, and transporting the animals to the show is to have them at their best at the time of showing. The final preparations to make them look their best include washing, filling, and bagging. The final washing should take place several hours before show time so the cattle have ample time to dry and eat. Many exhibitors prefer to wash their cattle the evening before the show. The timing of the last milking before the show should be determined by the milking interval at which the animal's udder looks best. Good showmen carefully observe the udders of show animals at various intervals after milking for several days before the show. Some may look best eight hours after milking, while others may look best with 14 to 16 hours of milk in their udders. Individual quarters on some cows may not be completely milked out, in order to balance the udder. It is very important to determine the approximate time a particular class will be shown and then perform the last preshow milking at such a time that the correct amount of milk will be in the udder at show time. Overbagging to an extent that the udder and teats have a distended appearance should be discouraged.

Most animals look better when they are full. They give the appearance of possessing more depth and width of body and are often more contented. A good fill may also serve to strengthen a slightly weak loin. Most showmen prefer to feed liberal amounts of hay, but to withhold water, concentrates, and succulent feeds, such as soaked beet pulp, for a period starting 12 to 15 hours before and extending to three to four hours before show time. This helps ensure a good appetite during the few hours just before the show, when the cows are fed succulents and concentrates. Water may be limited or withheld completely until just before showing. Care should be taken, however, to avoid overfilling cattle, as this may cause them great discomfort accompanied by roaching and difficult handling during the show.

Showing

The exhibitor should have final brushing, cleaning, and haltering taken care of and be nearby when the class is called. He should lead his animal around the ring in a clockwise direction, holding the halter strap in his right hand, but walking backward with the strap in his left hand when near the judge. From the time the animal enters the ring until she leaves it should be the business of the showman to see that the animal is being exhibited to her best advantage at all times. The judge's directions should be followed promptly and courteously. Most animals look better when walking than when standing, so the showman should keep his animal moving whenever possible. However, he should stop his animal and pose her whenever requested by the judge and when he is in the placing line. Cows in milk normally look best when standing with the rear leg nearest the judge slightly ahead of the other rear leg and with the front feet about even and spaced widely apart. Heifers are normally posed with the front legs in the same position as cows, but with rear leg nearest the judge slightly behind the other rear leg. The best position for any animal, however, is the one that presents her to best advantage. This should be determined prior to actual showing and the animal should be trained to pose in that position.

Chapter 22 contains the PDCA fitting and showing score card. The section on showing explains very explicitly the procedures that should be followed.

When the placings have been made the animals should be taken from the ring. Griping and criticism of the judge or fellow exhibitors must be avoided, even though the placement may not have been to the liking of the exhibitor. To be a good showman, a person must exhibit good sportsmanship.

Showing out of the Ring

Breeders are appreciating more and more the value of showing animals when they are not in the ring. Often prospective buyers or other interested people attend the show, and it is important that such people have the opportunity to see the animals outside the ring. This is especially true of winning animals. It is usually profitable to have a courteous and knowledgeable attendant with the animals when visitors are about. Unless the animals are about to be shown, the attendant should be willing to remove the blanket and show them at any time.

Fitting and Showing Ethics

The Purebred Dairy Cattle Association (PDCA) Show Ring Code of Ethics was designed with the belief that it is in the best interests of all breeders of registered dairy cattle to maintain a reputation of integrity and to present a wholesome and progressive image in the show ring. It therefore published a list of practices and procedures that are considered unethical in the show ring and should be avoided by all dairy showmen.

This code of ethics is presented at the end of the chapter.

Learning to Fit and Show Dairy Cattle

Basic principles of fitting and showing dairy cattle can be learned from discussions such as in this chapter. Real proficiency can only be acquired through actual experience working with cattle and by carefully observing the techniques of superior showmen. When attending fairs and shows as an exhibitor or a spectator, careful observation of the real professionals can be a valuable learning experience and is highly recommended.

Preparing Cattle for Sale

As the purposes of preparing cattle for sale are the same as for showing cattle—that is, to present them to best advantage at the time of the sale—similar procedures should be followed in regard to clipping, training, hauling, and so on. Consignors to sales should also be with their cattle while they are on display in the sales barn to promote their consignments and to answer questions from prospective buyers. All too often, consignors rely entirely on the sales team not only to sell their cattle, but to wash, clip, and train them to lead.

Sellers of dairy cattle should also be aware of the conditions of the PDCA Code of Ethical Sales Practices and strictly adhere to them.

THE PUREBRED DAIRY CATTLE ASSOCIATION SHOW RING CODE OF ETHICS[2]

The showing of registered dairy cattle is an important part of the promotion, merchandising, and breeding program of many breeders. Additionally, it is an important part of the program of the Purebred Dairy Cattle Association to stimulate and sustain interest in breeding registered dairy cattle. This relates to both spectators and exhibitors. In this connection, the PDCA believes that it is in the best interests of the breeders of registered dairy cattle to maintain a reputation of integrity and to present a wholesome and progressive image of their cattle in the show ring. It recognizes that there are certain practices in the proper care and management of dairy cattle, which are necessary in the course of moving dairy cattle to and between shows, that are advisable to keep them in a sound, healthy condition so that they might be presented in the show ring in a natural, normal appearance and condition. Conversely, it recognizes certain practices in the cataloging, handling, and presentation of cattle in the show ring which are unacceptable.

[2]Courtesy Purebred Dairy Cattle Association. This Code of Ethics was adopted by PDCA in March 1970 and was revised in February 1977.

The following practices or procedures are considered unacceptable and defined as being unethical in the showing of registered dairy cattle:

1. Misrepresenting the age and/or milking status of the animal for the class in which it is shown. In any female classes, animals may not be exhibited that are in milk due to an unnaturally induced lactation.
2. Balancing the udder by any means other than by leaving naturally produced milk in any or all quarters.
3. Setting the teats with a mechanical contrivance or with the use of a chemical preparation.
4. Treating or massaging any part of the animal's body, particularly the udder, internally or externally with an irritant, counterirritant, or other substance to temporarily improve conformation or produce unnatural animation.
5. Minimizing the effects of crampiness by feeding or injecting drugs, depressants or applying packs or using any artificial contrivance or therapeutic treatment excepting normal exercise.
6. Blocking the nerves to the foot to prevent limping by injecting drugs.
7. Striking the animal to cause swelling in a depressed area.
8. Surgery of any kind performed to change the natural contour or appearance of the animal's body, hide, or hair. Not included is the removal of warts, teats, and horns, clipping and dressing of hair, and trimming of hooves.
9. Insertion of foreign material under the skin.
10. Changing the color of hair at any point, spot, or area on the animal's body.
11. The use of alcoholic beverages in the feed or administered as a drench.
12. Administration of a drug of any kind or description internally or externally prior to entering the show ring, except for treating a recognized disease or injury and for tranquilizing bulls that may otherwise be dangerous or females in heat. For the purpose of this Code, the term "drug" shall mean any substance, the sale, possession, or use of which is controlled by license under federal, state, or local laws or regulations and any substance commonly used by the medical or veterinary professions which affect the circulatory, respiratory, or central nervous system of a cow.
13. Criticizing or interfering with the judge, show management, or other exhibitors while in the show ring or other conduct detrimental to the breed or show.

In keeping with the basic philosophy of the Purebred Dairy Cattle Association, ethics are an individual responsibility of the owner of each animal shown. Violations of this Code are subject to the disciplinary provisions of the appropriate dairy breed association and/or show management.

22

JUDGING OF FITTING
AND SHOWMANSHIP

The fitting and showing contests are such an integral part of a junior show that almost every judge will have an opportunity to officiate at many of these competitive events which demonstrate skill at preparation and exhibition of show animals.

One of the reasons for the popularity of this contest among the juniors at a show is that it stimulates interest in developing superior skill in the management of cattle and displaying them to best advantage. Also, it demonstrates to these young people the importance of doing the best possible job of raising, training, and fitting an animal before show day. Attendance at several spirited contests is almost certain to instill love and appreciation of good livestock, as well as tolerance and good sportsmanship.

The poise, skill, attractive personal appearance, responsiveness to directions, courage, courtesy, ability to remain cool and collected under pressure— all of which are needed for participation in the ring in these showing contests— are inspiring, character-building goals for these young people.

Furthermore, in addition to encouraging good animal husbandry, these contests point the way to good feeding and management, proper handling, constructive plans for improvement of type, and the art of good salesmanship, especially preparing and presenting animals to make them attractive to a prospective customer of dairy products or breeding stock.

To develop a standard judging program and to provide a guide for those who participate in fitting and showmanship contests, the Purebred Dairy Cattle Association developed, to the satisfaction of the American Dairy Science Association, a score card for fitting and showmanship contests. Appreciation is expressed for their permission to reprint this information as follows:

UNIFORM SCORE CARD FOR JUDGING FITTING AND SHOWMANSHIP CONTESTS

A. Appearance of Animal—40 points
1. Condition and thriftiness, showing normal growth, being neither too fat nor too thin—10 points
2. Grooming—10 points
   a. Hair properly groomed and the hide soft and pliable. Hair dresser should not be used in excess.
   b. Hooves trimmed and shaped to enable animal to walk and stand naturally.
   c. Horns (if present) scraped and polished.
3. Clipping—10 points
   a. The final clipping should be done about two days before show.
   b. Head, ears, tail, udder, and elsewhere clipped as needed but not over entire body. Belly and udder not to be clipped on heifers that have not freshened and are not springing close.
4. Cleanliness—10 points
   a. Hair and switch clean and if possible free from stains.
   b. Hide and ears free of dirt, and legs and feet clean.

B. Appearance of Exhibitor—10 points
1. Clothes and person neat and clean; white costume preferred.

C. Showing Animal in the Ring—50 points
1. Leading—15 points
   a. Enter leading the animal at normal walk around the ring in a clockwise direction, walking opposite her head on the left side, holding a lead strap (or rope) with the right hand quite close to the halter with the strap neatly, but naturally (not necessarily coiled), gathered in one or both hands. Holding close to the halter ensures a more secure control of an animal.
   b. Animal should lead readily and respond quickly.
   c. Halter of right type, fitting properly and correctly placed on animal. A leather halter with leather strap is best.
   d. As the judge studies your animal the preferred method of leading is walking slowly backward facing the animal and holding the lead rope in the left hand with the remainder of it neatly, but naturally, gathered in one or both hands (face forward when leading at all other times).

e. Lead slowly with the animal's head held high enough for impressive style, attractive carriage, and graceful walk.

2. Posing—15 points
   a. When posing and showing an animal stay on the animal's left side and stand faced at an angle to her in a position far enough away to see stance of her feet and her topline.
   b. Pose animal with feet placed squarely under her with the hind leg nearest to the judge slightly behind the other one when showing heifers. The hind leg nearest the judge should be slightly ahead of the other one when showing cows.
   c. Face animal up-grade, if possible, with her front feet on a slight incline.
   d. Neither crowd the exhibitor next to you nor leave enough space for another animal when you lead into a side-by-side position.
   e. Animal may be backed out of line when judge requests that her placing be changed. Many prefer to lead animal forward and around the end of the line or back through the line. Do not lead animal between the judge and an animal he is observing.
   f. Do most of the showing with the halter lead strap and avoid stepping on animal's hind feet to move them.
   g. Step animal ahead by a slight pull on the lead strap.
   h. Move animal back by exerting pressure on the shoulder point with the thumb and finger of the right hand as you push back with the halter.
   i. When judge is observing the animal, let her stand when posed reasonably well.
   j. Be natural. Overshowing, undue fussing and maneuvering are objectionable.

3. Show animal to best advantage—10 points
   a. Quickly recognize the conformation faults of the animal you are leading and show her to overcome them. You may be asked to exchange with another and show her or his heifer for awhile.

4. Poise, alertness, and attitude—10 points
   a. Keep an eye on your animal and be aware of the position of the judge at all times. Do not be distracted by persons and things outside the ring.
   b. Show animal at all times and not yourself.
   c. Respond rapidly to requests from the judge and officials.
   d. Be courteous and sportsmanlike at all times.
   e. Keep showing until the entire class has been placed and the judge has given his reasons.

A summary of the main points in condensed form follows:

Points

Appearance of Animal 40

Condition	10
Grooming	10
Clipping	10
Cleanliness	10

Appearance of Exhibitor 10

Showing Animal in the Ring 50

Leading	15
Posing	15
Show animal to best advantage	10
Poise, alertness, attitude	10

100

Before this score card was available, a number of controversial points resulted in a certain amount of confusion among the exhibitors and in some lack of uniformity in agreement among the judges.

Participants who study these directions carefully and follow them before and during the contest should do an outstanding job and place high in competition with others. The judge who uses this information as a guide and works diligently to make an accurate appraisal can make placings that will be above criticism and meet with a general agreement and approval from the contestants and spectators.

23
YOUTH JUDGING
(4-H, FFA, and Other Organized Groups)

Dairy cattle judging is an excellent incentive for young people to learn more about dairy cattle and the dairy industry. Through judging, young people visit many farms and cattle shows, where they learn much about the breeding and management of dairy cattle and meet many persons with similar interests. Judging improves their ability to make logical decisions with supportive reasoning and to express their decisions in a clear, well-organized manner. These experiences help youths to better understand and meet the future challenges of the dairy industry and life. Maurice Mix, a 4-H member of a winning New York team in national competition, said, "In all probability I would not have become director of international affairs for the Holstein Association had it not been for my 4-H livestock and dairy judging training, my national contest participation, educational trip experiences, and the enthusiasm, motivation, and the personal interest and encouragement of 4-H Extension personnel."

STEPS IN LEARNING TO JUDGE

In learning dairy cattle judging, one needs to learn the names and locations of the parts of the dairy animal. This is very important in order to correctly observe, evaluate, and compare dairy cattle. A good working knowledge of body parts and functions enables youngsters to express themselves more clearly in both written and oral reasons. The body parts are illustrated in Chapter 2.

Along with learning the body parts, youths should learn dairy animal comparative terms. A comprehensive vocabulary enhances one's ability to understand, organize, and clearly state the reasoning for placing a class of animals. Each individual should develop terms that relate to dairy character, general appearance, body capacity, and mammary system. A list of some comparative terminology is included in Chapter 14 and other chapters throughout this book.

In youth judging, one should first learn and use basic comparative terms. With experience, the vocabulary can be expanded with more precision. Chapters in this book pertaining to body conformation and relevant descriptive terminology should be studied. After studying the chapters, one should select the terms to be used in reasons. The learning of body parts and comparative terms is essential in dairy cattle judging. An activity similar to dairy quiz bowls may be beneficial and challenging in learning about judging.

Dairy cattle judging involves the comparison of each animal to the ideal reference animal and to the animals within a class. Young people should know what the ideal reference animal is for each breed and age. This can be accomplished by studying pictures in this book and in breed journals, and remembering the good-quality cattle from state and national shows. Youths should develop a clear vision of the ideal reference animal for each breed and age in order to be confident in judging all breeds.

Before accurately judging dairy cattle, youths need to study the score card for dairy cows. A detailed discussion of its use appears in Chapter 3 along with its relative assignment of points for general appearance, dairy character, body capacity, and mammary system. For heifers, general appearance should be allotted approximately 55 to 60 points, dairy character 20 to 25 points, and body capacity 15 to 20 points. In youth judging, the mammary system (the promise of) is not judged, because only normal heifers without a deficiency are selected for classes. For practical application on udder promise in heifers, see Chapter 18.

In judging dairy cattle, one should develop a systematic approach that lends itself to an organized sequence of analysis. This will maximize organization, use of time, and logical thinking in judging dairy cattle. At the beginning of each class, one may evaluate the strengths and weaknesses of each animal according to the ideal reference animal and then compare the animals in the class to each other. This should be done at a distance so all the animals can be observed at the same time. It is a common mistake to judge the animals from close up, where all the animals cannot be seen. After evaluating and comparing the animals, a tentative placing should be made by mentally listing the major points of differences between animals in each pair. Once you have decided on a tentative placing, you should quickly justify the placing in your mind before making the final placing. This will minimize making placings that are not logical and are made without confidence.

Reasons are essential in justifying the placing of a class. It is beneficial for young people to develop and use a format for organizing and delivering a set of reasons. One should study Chapter 14, which relates to reasons for placing of classes.

ORGANIZATION OF REASONS

The following are suggestions for organizing a set of reasons.

1. In the introduction, you should state the breed and age along with the placing. Following this, a brief statement should be made pertaining to the analysis of the class.
2. Within each pair, comparative terminology should be used in justifying the placing of one animal over another. Grants should be stated when major differences exist in favor of the second-placed animal over the first-placed animal in the same pair. Grants should be brief and comparative.
3. In the closing statement, a brief statement should be made pertaining to the major reasons for placing the last-placed animal at the bottom of the class. Following this statement, one may state a positive strength(s) of the last-placed animal, thus ending the reasons on a positive note.

DELIVERY OF REASONS

1. Stand erect on both feet, approximately 8–10 feet from the judge. Put hands at your side or behind you. Avoid moving hands and feet needlessly. However, a few hand gestures may be used to reinforce the delivery.
2. Make eye contact with the judge prior to giving the reasons. Maintain eye contact throughout the reasons and questions. You may wish to look just above the judge's head; however, you must give the impression that you are looking at him or her.
3. You have two minutes to deliver the reasons. Practice delivering reasons within this time.
4. Organize the reasons so major points of differences within a pair are given first. Minor differences should be included only if they are of importance. Avoid minor differences that are questionable and unimportant.
5. Do not memorize reasons. When giving reasons, visualize the animals in pairs as they were placed. With practice, you will gradually develop the technique of recalling the animals in a given class and applying the correct comparative terminology while giving the reasons.

The following example of reasons is given to demonstrate the type of format, organization, and comparative terminology that may be used to write or orally present an excellent set of reasons in justifying a placing of a class.

EXAMPLE SET OF ORAL REASONS

I placed this class of Guernsey 4-year-olds 4–3–2–1. I found an outstanding top in the best-uddered 4; that placed over the good topline 3; followed by the dairy 2; and finished the class with the least dairy 1.

I placed 4 over 3 because she excels in size and scale, being taller at the withers, longer through the barrel and rump, and deeper and wider in both fore and rear rib. Four has an advantage in dairy character, being cleaner through the head and neck, sharper at the withers, more open in the ribs, and cleaner in the thighs. Furthermore, 4 has a smoother and firmer fore udder attachment, more correct teat size and shape, and a higher rear udder attachment. I grant 3 has a more desirable set to her rear legs.

I placed 3 over 2 because she has more general appearance, being straighter and smoother over the topline, fuller in the crops, and neater in the tail-head setting. Three is also higher in the rear udder and more nearly level on the udder floor. In addition, 3 is straighter in the rear legs, showing less hocking from the rear. Three is wider in the chest floor, and deeper and wider in the fore rib. I admit 2 shows more dairy character, especially through the front end.

I placed 2 over 1 because she has a definite advantage in dairy character, being cleaner and longer in the head and neck; sharper over the withers; more open in the ribs; and cleaner about the hooks, pins, and thighs. Two is taller, and has an advantage in her mammary system, showing a more desirable curvature to the rear udder and having a more defined median suspensory ligament. I reconize 1 is fuller in the crops and tighter at the point of elbow.

I placed 1 at the bottom because she lacks the dairy character and stature of the cows above her, but I admire her for the width of chest floor and levelness of rump. For these reasons I placed this class of Guernsey 4-year-olds 4–3–2–1.

High Individual on Placings, Second on Reasons, and High Individual in Contest on a High Team at the National 4-H Contest.
Matthew Budine,
Walton, New York

Figure 400   Matthew Budine, Walton, New York

24

USEFUL RULES
and
GRADING OF PLACINGS FOR
COMPETITIVE JUDGING

Effective guiding principles are the key to almost all successful endeavors. Some rules to serve as a guide to crystallize thinking and judgment are especially useful to the young judge, who cannot be expected to have the self-assurance and confidence of an experienced veteran of many judging assignments.

As an illustration of the effectiveness of the rules for competitive judging, the senior author has in mind a college student with limited experience and background who was once judging the difficult classes in the National Intercollegiate judging contest. Several times during the contest this student could not get the classes to "unfold" for him. Nearly every judge has at one time or another had a similar experience. This student solved the problem by turning his back on the classes, calling to mind the ideal types, and asking himself what he should be looking for in each particular class. In almost every instance he found the correct answer, and he finally ranked near the top in the contest for judging all breeds.

USEFUL RULES TO GUIDE PLACINGS

The rules presented in this chapter have on many occasions guided student judges in reaching a sound decision in competitive and practice judging. The professional judge, likewise, will occasionally find these useful in show-ring judging. They offer recommendations for procedure in difficult situations, and thus help to prevent confusion when a problem appears particularly formidable.

The following 15 rules can serve as a guide to maintaining a reasonable degree of uniformity for judging variable type characteristics:

1. The first-place individual should be well balanced, smooth, symmetrical, and of proper size, should possess an outstanding udder and dairy character, should have strong legs, and should be free from any major defects.

2. The bottom-place individual is the one that appears the most unattractive, is unbalanced and lacking in symmetry, and has one or more major defects.

3. Larger animals place over smaller ones if both are alike in points of conformation. In other words, size is usually considered an advantage if the animal is not coarse or too large. Quality with size is important.

4. Small but smooth, well-balanced animals are usually placed over larger animals that have major defects, or over larger animals that lack smoothness and symmetry or a good udder. In other words, a small, good animal that has plenty of quality is preferred to a large one that can be criticized on a number of points.

5. In cow classes it is a safe rule, for close placings, to choose the cow that has the best udder. Cows placed toward the top of the class should always have good udders.

6. A broken-away or pendulous udder puts the cow at the bottom of the class, regardless of other good points of conformation. If two or more animals in the class have broken-away or pendulous udders, they are then placed on points of body conformation, but this is the only exception to the rule. It should be well established, however, that the udder is pendulous; this can be determined to a large extent by the next rule.

7. If the floor of the udder where the teats attach hangs lower than the point of the hock on the cow's leg, the udder is usually considered pendulous. This is also a good rule to determine whether or not an udder hangs too low. Usually, the rule holds true even immediately before freshening, but good judgment (after careful study) has to be used at this time of gestation.

8. There are various degrees of deviation for any point that is under consideration. The advantage one cow is given over another, or the amount of discrimination placed on a certain point, depends directly on the degree or extent of the defect. Another point that enters here is the relative importance of the part that is defective; this can be determined from the Dairy Cattle Judging Score Card.

9. An animal good or outstanding in all points but one, if this point is not too important (this does not include the udder), can usually be placed in second position with strong justification. Examples of such points are a high tail head, heavy withers, easy loin, and other similar points. Assign-

ing such animals to second place indicates that even though the good qualities are appreciated, the animal cannot be placed at the top where the deficiency would receive criticism or objection.

10. The legs should be observed while the animal is both in motion and at rest. The strength and set of the legs can best be noted while the individual is moving. The strength of the fore udder attachment can be studied at the same time, and a swinging udder caused by a loose fore attachment or loose medial suspensory ligament can easily be detected.

11. Each animal should be observed several times to determine whether or not the individual settles when standing in one position. Weaknesses that may have been overlooked previously can definitely influence the placings.

12. In general, the same points of judging, except breed type, apply to all breeds. Brown Swiss and Holsteins are just about the same, except for color, a slight difference in depth of body in favor of the Holsteins, and ruggedness in favor of the Swiss. Refinement and size are important for Guernseys and Jerseys.

13. The ideal type and the ideal conformation for each specific point should be kept in mind, because this will make it easier to select cows (or bulls) for the top placing that most nearly conform to these ideals. A careful analysis of each individual will expose deviations. One should always remember that, for cow classes, individuals toward the top of the class must have good udders and dairy quality.

14. The points that should always be given careful consideration and that are most important in judging are udder, dairy character, legs, topline, size, general appearance, good breed type, smoothness, and general blending of parts throughout the body.

15. The cows that should be selected are those that have outstanding dairy character and general good type and that appear to be the useful, hard-working kind that will wear well with age, when conformation differences begin to resolve themselves into economic values. An individual should be evaluated from the standpoint of practical usefulness, with special emphasis on udder, legs and feet, dairy quality, stature, and strength.

GRADING OF PLACINGS

The grading of placings on results both for judging under competitive conditions and for classroom work have passed through various evolutionary stages but are now standardized on a sound basis. At one time, standard cuts were used regardless of the closeness for placing a pair or a class. This procedure was unscientific, and gradually a system was developed whereby the recommended cuts are based on assigned values between the various pairs of animals for each class. These assigned values are based on the degree of difference between two individuals for which the cut is assigned.

Grading Procedure

Grading for placing can be on the basis of either 50 or 100 for a total score. The basis of 100 has an advantage for classroom work since it permits a direct average of the score for assigning grades in a course. The score of 50 for a perfect placing is preferred for competitive contests because if reasons are graded on the same basis, it will provide 100 points as a total for the placing and the reason grade on a class.

When the grading is on the basis of 100, a maximum cut of 30 will give a score of 0 on a reverse placing. If the grading is on the basis of 50, a maximum total or additive cut of 15 will give a score of 0 on a reverse placing. If five to ten individuals are in a class, as for use in breed and official judges' conferences, the cuts must be correspondingly lower; often there should be only a one-point cut between a close pair. If the deductions or cuts are high and the placings difficult for large classes, it is possible to get a minus score. In grading such placings, the actual score assigned should not be less than zero. This avoids overemphasizing one class in calculating the total score.

Tabulated score cards are available and can be commercially purchased. An example on the basis of 100 as a perfect score is cited on page 322 to show how deductions are made when the official placing is 3–4–2–1, with cuts of 8, 10, and 4 between the respective pairs.

The example demonstrates how deductions are determined from various switches in placings, with the cuts adjusted according to how obvious the placing appears to the official judge. Only nine different placings of the 24 possible ones are presented. After some practice, it is relatively easy to sort the cards by different placings and assign the proper score to each.

When the official placing is made for competitive or practice judging, each ring is assigned individual cuts between the respective animals. These cuts are designated by the official judge in accordance with the extent of the differences that exist in the placings between the animals.

Some prefer to do this systematically, and in many contests it is advisable to keep a record of the score assigned to each placing. In such instances, it is possible to use a special grade sheet (shown at the end of the chapter). The formulas on which deductions are based shorten the time required and increase the accuracy of determining the grade for each of the 24 possible placings.

The sample at the end of this chapter is reproduced with the permission of Dr. R. Neidermeier, Department of Dairy Science, University of Wisconsin. He uses a separate page for each of the 24 official placings possible with a class of four animals. Space does not permit reproduction in this book of each of these sheets. However, the form is the same for each sheet except for the *Plac-*

ing column; this column is presented according to the numbers and formulas, from 1 through 24, respectively. With this information, the reader can make up sets of mimeographed sheets for each of the 24 placings possible in a class of four animals. After these grading sheets have been made up, one will greatly appreciate the tabulators commercially available for this work. These are carefully checked and entirely accurate. Computers are now frequently used to grade the results from judging contests.

Although grading judging cards is time-consuming and often tedious, it is an important part of the work. One should recheck the grades on all cards and also the calculations on the tabulating or record sheet as an insurance against mistakes. Accuracy is very important in fairness to each participant of both competitive and practice judging.

EXAMPLE FOR DEDUCTIONS

	Cuts 8 10 4	
Official Placing:	3—4—2—1	Score—100
	3—4—1—2	Score— 96
	4—3—2—1	Score— 92
	3—2—4—1	Score— 90
	4—3—1—2	Score— 88
	2—3—4—1	Score— 72
	2 over 4 deduct	10 points
	2 over 3 deduct	18 points
	Total deduction	28 points
Placing: 1—3—4—2		Score—60
	1 over 2 deduct	4 points
	1 over 4 deduct	14 points
	1 over 3 deduct	22 points
	Total deduction	40 points—Score 60
Placing: 1—2—3—4		Score—32
	To the 40 point deduction above add	
	2 over 4 deduct	10 points
	2 over 3 deduct	18 points
	Total deduction	68 points—Score 32
Placing: 1—2—4—3		Score—24
	4 over 3 deduct	8 points
	2 over 3 deduct	18 points
	1 over 3 deduct	22 points
	2 over 4 deduct	10 points
	1 over 4 deduct	14 points
	1 over 2 deduct	4 points
	Total deduction	76 points—Score 24

Ring ....   Breed ....

GRADES FOR OFFICIAL PLACING—3 4 2 1
        Cut   **8 10 4**
Official Placing   **3—4—2—1**

Multi-plier	Top	Middle	Bottom
	t	m	b
1	8	10	4
2	16	20	8
3	24	30	12
4	—	40	—

			Cut		*Placing*				*Grade*
1.	100—none	or	....	=	3	4	2	1	....
2.	100—(b)	or	....	=	3	4	1	2	....
3.	100—(m + 2b)	or	....	=	3	1	4	2	....
4.	100—(2m + 2b)	or	....	=	3	1	2	4	....
5.	100—(m)	or	....	=	3	2	4	1	....
6.	100—(2m + b)	or	....	=	3	2	1	4	....
7.	100—(t)	or	....	=	4	3	2	1	....
8.	100—(t + b)	or	....	=	4	3	1	2	....
9.	100—(2t + m)	or	....	=	4	2	3	1	....
10.	100—(3t + 2m + b)	or	....	=	4	2	1	3	....
11.	100—(2t + m + 2b)	or	....	=	4	1	3	2	....
12.	100—(3t + 2m + 2b)	or	....	=	4	1	2	3	....
13.	100—(t + 2m)	or	....	=	2	3	4	1	....
14.	100—(t + 3m + b)	or	....	=	2	3	1	4	....
15.	100—(2t + 2m)	or	....	=	2	4	3	1	....
16.	100—(3t + 3m + b)	or	....	=	2	4	1	3	....
17.	100—(2t + 4m + 2b)	or	....	=	2	1	3	4	....
18.	100—(3t + 4m +2b)	or	....	=	2	1	4	3	....
19.	100—(t + 2m + 3b)	or	....	=	1	3	4	2	....
20.	100—(t + 3m + 3b)	or	....	=	1	3	2	4	....
21.	100—(2t + 2m + 3b)	or	....	=	1	4	3	2	....
22.	100—(3t + 3m + 3b)	or	....	=	1	4	2	3	....
23.	100—(2t + 4m +3b)	or	....	=	1	2	3	4	....
24.	100—(3t + 4m + 3b)	or	....	=	1	2	4	3	....

25
PRACTICAL IMPORTANCE OF TYPE AND HERD CLASSIFICATION

Every discussion on type should start with the following statement: "Production is basic and must receive first consideration; however, functional type must receive proper attention in order to capitalize on the high production bred into the modern dairy cow." High production and good type for wearability will dictate how the dairy cow can compete in the country's food market. Good type but with poor production is relatively useless.

A dairy cow needs type for commercial adaptation. You should keep in mind a cow's wearability and workability when you have to make a living with her. A cow may have remarkable production for one or two lactations, but if, for example, her udder breaks down on the second or third lactation because of a weak suspensory ligament, she is almost worthless. She barely reached the break-even point on the cost of raising a heifer (Figs. 401 through 406). Similarly, a young cow that has bad legs at the age of 4 or 5 years contributes little to a successful dairy enterprise (Fig. 408). For the dairy farmer or breeder to stay in business, he or she needs functional type for wearability. The cows should also have good workability traits so that the dairy farmer can enjoy working with the herd.

A dairy farmer and (or) breeder who appreciates good type in cattle often provides improved herd management and good husbandry. This enables him or her to keep good, sound cows around year after year, which in turn frees

Figure 401   An outstanding Holstein cow that is the practical kind. She has all good points on conformation and no weaknesses, and she is the kind that will be in the herd for a long time. Note her dairy quality with strength; she is a powerful cow without any sign of coarseness. Her well-attached udder, both fore and rear, is carried by a medial suspensory ligament that is at its best in this high producer (in 305 days, 28,240 lb of milk and 1175 lb of fat). She stands on a good set of legs and her feet are superb with a deep heel and a well-shaped foot. She has never had any foot trouble. All of these good traits were recognized when this cow classified 3E 93 (3X Excellent in required age brackets with a final score of 93). (Courtesy Carnation Farm, Seattle, WA)

other well-bred replacements for sale; or he or she may do additional culling for greater selection pressure on production. Some dairy farmers and (or) breeders have tried to advance the idea that a rapid turnover takes advantage of genetic progress. This is a shallow argument because at least the top half of the herd is years ahead in what can be added on production in one generation.

The successful dairy farmer and (or) breeder emphasizes both production and type (Figs. 401, 403, 404, 409, 411, 412, 417, 418, 419, 420, and 421). He or she keeps these in proper perspective and has a program for breed improvement far beyond the basic requirement of high production. The aim is profitable milk production under conditions of optimum management. Both purebred breeders and commercial dairy farmers stress type traits that are associated with wearability, workability, salability, and profitability. These characteristics are listed as follows:

Wearability:

1. Strong udder support—a strong medial suspensory ligament with clear halving on the floor of the udder.
2. A high, wide, firmly attached rear udder.
3. A fore udder of moderate length and firmly attached.
4. Teats that hang plumb, are of desirable size and length, and are squarely placed.
5. Correct rear feet and legs; moderate set to hocks and sound feet.
6. Dairy character, which is representative of a cow's milking ability. Thick, overconditioned young heifers have too much fat in their udders and they generally produce less. Older cows that are overconditioned have lower production, require more maintenance, are more prone to injury, and show greater stress at calving time, including susceptibility to displaced abomasa and ketosis.

Conversely, some type characteristics found to have a negative effect on wearability and lifetime production are:

1. Coarse front ends.
2. Crampiness.
3. Broken udder support.
4. Large teats.
5. Loose rear udder.
6. Long, wide, level rumps.

All of these characteristics are heritable, but to varying degrees. Reported values vary between breeds. Dairy character and stature are generally relatively high (0.35-0.45) while rear legs are relatively low (0.10-0.15). Heritability of final score varies between breeds from 0.23 in Jerseys to 0.43 in Brown Swiss, with the average being about 0.30. A major point is that these characteristics are heritable and we can make genetic progress in improving the traits associated with wearability if we can measure these traits fairly accurately.

Workability:

1. Milking speed.
2. Proper teat size, shape, and placement.
3. Even temperament or disposition.

4. Freedom from crampiness.

5. Intermediate stature or above because an upstanding cow produces cleaner milk, has less udder injury, and is more convenient to work with at milking time.

6. A good appetite and proper body capacity.

7. Mastitis resistance.

The above characteristics affect profitability because these cows can produce at high levels over long lifetimes without needing pampering or other special care. The commercial dairy farmer needs such cows, and, because cows have to last, he or she wants only those with good feet and legs, with strongly attached and well-supported udders of top quality, and with strong frames and bodies that possess dairy quality in order to stand the stress of high production.

Salability affects profitability in that the dairy farmer has the kind of cows that his or her neighbor wants to buy from his or her surplus stock or that have the size and strength to bring a good market price as cull cows. The practical producer believes that emphasis on type is motivated by maximum profit.

Figure 402 This is a strong-framed cow like the one shown in Fig. 401, but the medial suspensory ligament in her udder is broken. Because of this, she developed a meaty, swinging udder that is below the point of the hock and therefore is a pendulous udder. Because her lifetime in a practical herd is very limited, she will place last in a judging class unless there is another cow that has a pendulous udder, in which case both cows will be placed on other points of conformation. This cow is not as outstanding in dairy character as the one in Fig. 401 because of differences in cleanness up front in her head, neck, and shoulders. She is not as sharp over the back, the hips, or pins. Her feet and legs cannot compare because she has soft pasterns and a shallow heel on her foot that has given her a lot of trouble and has required considerable veterinary care. Under modern production methods, this type of cow does not last in a herd; therefore, she is not the economical kind. She was a moderately high producer for a relatively short time.

Figure 403 An outstanding young cow that was selected All-American at 2, 3, and 4 years of age. Later she was voted All-Time All-American 2-year-old, 3-year-old, and Reserve All-Time All-American 4-year-old. She classified excellent with a score of 97 at 5 years, 2 months. She is the youngest cow to receive the top score of 97. This cow is the useful kind, as verified by the following production records during corresponding years of the All-American awards before 5 years of age:

1 y - 10 mo	285 d	16,951 lb milk	3.9%	668 lb fat
2 y - 9 mo	311 d	23,135 lb milk	3.6%	855 lb fat
3 y - 9 mo	339 d	23,343 lb milk	3.4%	788 lb fat
4 y - 9 mo	327 d	18,461 lb milk	3.9%	719 lb fat
5 y - 9 mo	365 d	27,322 lb milk	3.5%	956 lb fat

Note the perfect regularity in reproduction. (Courtesy Bower Farms, Inc., Ray Vail, Manager, LaGrangeville, NY)

Figure 404 This udder is on an Ex 97 cow with an Excellent and all 1s on mammary. As a senior 5-year-old, this udder produced 33,471 lb of milk, 3.7%, and 1215 lb of fat. She was Reserve All-American during her first four lactations before she was All-American Aged Cow for 2 successive years. (Courtesy Roy Hetts, Fort Atkinson, WI)

Figure 405 The broken medial suspensory ligament on the udder of this cow permitted the floor of the udder to drop to the extent that it pushed the teats to the side of the udder. It made the cow worthless (except for beef) as a milker in a modern production pattern.

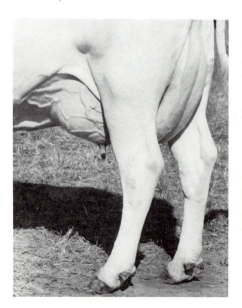

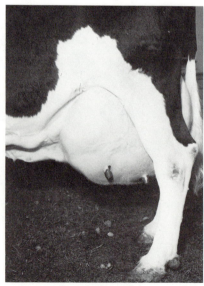

Figure 406 This udder has a broken fore and rear attachment. Also, the broken medial suspensory ligament in the center of the udder permitted the udder to drop down far below the point of the hock. It then became a very swinging, pendulous udder, an udder that is subject to injury. This and the position cause the tissues to become a very meaty udder. It is obvious that a dairy farmer has no place for this kind in an efficient milking herd. Unfortunately, there are still some cows like this and they require early replacement. Fortunately, because dairy farmers are using the information in the Sire Summaries, the number of individuals having this objectionable udder decreases each year.

Figure 407 This cow has almost perfect rear legs and feet. She stands on a strong, clean, flat bone which terminates in a well-formed foot. (Courtesy Hanover Hill Farm, Armenia, NY)

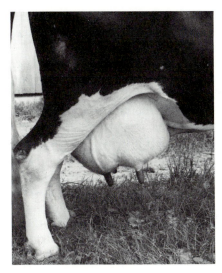

Figure 408 Broken-down legs and feet and broken fore udder attachment are causing this cow serious difficulty and will keep her from having a useful, highly productive life. This type needs pampering and frequent veterinary care. The owner will soon look for a replacement.

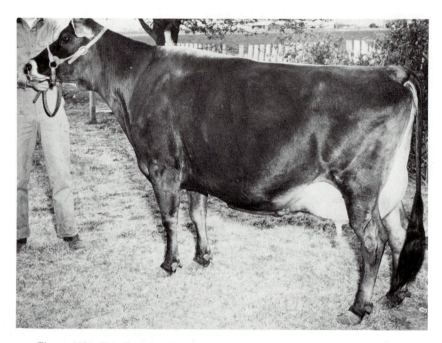

Figure 409 This Excellent Jersey excels in production and has a record of 18,410 lb of milk and 1008 lb. of fat actual as a 2-year-old and 22,360 lb of milk and 1235 lb of fat actual at 3 years. She produced another 1047 lb of fat at 6 years, 11 months of age. She is the only eighth-generation 1000-lb fat producer of any breed. She has a daughter that has a record of over 1000 lb of fat. (Courtesy American Jersey Club, Columbus, OH, and Victory Jersey Farm, Tulia, TX)

All breeders and commercial dairy farmers and all producers working between these two groups should set up their own goals for production and type. Each one must decide what is best for his or her situation and then select accordingly. A dairy farmer who changes from a conventional barn, pasture feeding, and individual cow management system to continuous confinement and group handling may wish to make changes in type specification. Fortunately, research has developed informational tools the dairy farmer can use to change the type of his or her cattle. The AI and the various breed organizations will also assist the dairy farmer in this endeavor.

The commercial dairy farmer, the purebred breeder, the AI organizations and membership, the breed association personnel, and many others fully appreciate what the modern cow needs in order to compete in a freestall operation, and they appreciate other modern forms of handling and of housing the milking herd. They realize the advantage of a cow that is unconcerned with the unusual in her surroundings but is an aggressive feeder with a businesslike competitiveness with other cows in the herd. Consultants for charting planned matings for the entire herd of a client, the AI studs with their special programs, and the breed associations with their unprecedented activity on type evaluation

Figure 410 The lack of dairy character in this low (11,000 lb of milk) producer is reflected in her coarseness about the head, thick neck, heavy shoulders, tight and close rib, and overall cover of flesh. The good judge detects this, but many judges are taken in by the strength, upstandingness, straight topline, well-attached udder, and satisfactory legs. A small percentage of cows with the traits displayed by this cow will be good producers, but not very many. Compare her to the features for dairy character in the cow in Fig. 411.

Figure 411 This sharp, lean, angular cow with a refined head and neck, sharp shoulder, hips, and pins with well-defined vertebrae in her back indicates dairy character which reflects her production of over 16,000 lb of milk as a 3-year-old. Compare her with the cow in Fig. 410 and note the opposites on points that denote dairy character. Rarely will the sharp cow showing dairy character displayed by this cow be unsatisfactory in production. (Courtesy Merton Sowerby, Woodacres Farm, Princeton, NJ)

have all had a tremendous impact on the improvement of dairy cattle. From this, good dairy farmers have quickly learned to turn their backs on bulls that have passed on serious type weaknesses to their progeny.

The Sire Performance Summaries computed by the U.S. Department of Agriculture and published by the Extension Service and Breed Associations have provided extensive and reliable information on production and type for the progeny of a bull. This information and the wide distribution of published results have provided opportunities for improvements that have never been available before in the history of dairy cattle breeding. Because of this, terms like *trait mating, type appraisal results, descriptive type traits, mating guide, computer matching, genetic engineers' mating guide, complete cow program, the eye of the master, PD* (predicted difference), *PDT* (PD type), *PD$* (PD dollars), *PTI* (production type index), *planned breeding,* and *California Physical Traits Program* have become an active part of the program and serve a useful purpose to the practical dairy farmer.

Several of the purebred associations now offer combined programs of production and type evaluation. Most use the descriptive type traits evaluation program (or one very similar) developed by a special committee of the National Association of Artificial Breeders (NAAB). Some breeders also instruct their classifiers to assign an overall numerical score and major component numeri-

Figure 412 A useful kind of Ayrshire cow in a practical dairy farmer's herd. This cow was outstanding for production, was successful in the show ring (All-American 4-year-old and Aged Cow), and displays excellent type for herd classification. (Courtesy Haynes Ayrshire Farm, Tully, NY)

cal scores. It is the fervent hope of all associated with the dairy industry that this latest revision and use of the new classification system will provide more accurate information to make more rapid genetic progress in improving type in dairy cattle. Research has demonstrated that type is important to commercial dairy farmers and that many components of type contribute to and others detract from longevity and lifetime production. In addition, type contributes almost one-half of the selling price of purebred animals.

TYPE CLASSIFICATION

In its early stages of development, type classification was an attempt to evaluate more accurately the physical conformation of dairy animals. Basically, the animal to be classified was compared to the ideal of the breed and then was assigned a score on major components of type (dairy character, body capacity, and so forth). A perfect score was 100. Various type classification brackets were used: Excellent for scores of 90 or higher; Very Good, 85 to 89; Good Plus or Desirable, 80 to 84; and so on down through Good or Acceptable, Fair, and Poor. Theoretically, a cow that classified Very Good with a score of 86 points of a possible 100 was similar in body conformation to other 86-point cows of the breed regardless of herd or location. This was a more meaningful evaluation of her type than a statement that she stood third in her class at the District Dairy Show.

Shortcomings of the system included the following: different score cards for different breeds; inclusion of purebreds only, and many biological traits included in the same type component. For example, a cow might score poorly in general appearance because she had poor legs but was outstanding in topline, rump, and shoulders. Regardless of these problems, it was a beginning in evaluating type more accurately than in the show ring where each animal was compared to every other animal in that class that day. In this case, a cow very poor in type might place first because of lack of competition or a very fine cow might place tenth because of very superior competition. There were also similarities between classification and show-ring performance. In classification, as in the show ring, the emphasis is on type characteristics or conformation necessary for a long, useful life and sustained high production, a low incidence of udder disturbances, and freedom from sore feet, bad legs, and other conditions that increase veterinary costs or affect the economic value of a cow. To effect the desired type characteristics in an animal, the most emphasis should be placed on udder conformation, dairy character, feet and leg structures, muscle tone, constitution, and general strength, together with smoothness of parts, as evaluated in the section on general appearance.

The skilled and experienced official type-classifier recognizes that there is a right and wrong body shape for proper dairy type. He or she discriminates against a cow that has a heavy, compact, round body because this indicates that the animal converts a larger amount of feed into body maintenance and fat than does a cow that has more dairy quality. He or she discriminates against a very frail cow because she hasn't the strength or constitution to stand the rigorous requirement for high production year after year.

Usually there is general and close agreement between show-ring placings and official type-ratings, but there are numerous and well-justified exceptions. For type classification, the emphasis is entirely on the way the cow appears in the herd from the standpoint of production and breeding purposes. Unbalanced quarters and other notable deficiencies are evaluated differently for the two purposes. For example, a blind quarter in the show-ring is a disqualification, but for classification this receives only minor criticism if due to an injury or infection. Likewise, a side leak on the teat is evaluated much differently in the show ring than during classification.

In the show ring a cow of recent freshening is often given more of an advantage over a stale cow toward the end of her lactation than is true for classification. Body changes due to age are often considered from a more favorable viewpoint in classification than for show-ring purposes, where the emphasis is on utility factors *plus* beauty, charm, and gracefulness, for which the advantage always rests with youth.

The original classification programs proved to be an excellent tool for unbiased evaluation of the type in purebred herds. They pointed out weaknesses and strengths of individuals and groups of animals from a particular sire or cow family. They were also an excellent merchandising tool, especially for excellent and very good animals. These type ratings added considerably to the price when selling these animals or offspring from these animals.

There was some variation in classification score due to the following: (1) age (often younger cows were appraised rather low with the justification that they could be raised at a later date); (2) stage of lactation (fresher cows often had a slight advantage over stale or dry cows); (3) sick or injured cows; and (4) marked changes in conformation, especially breakdown of udder, legs, and topline, and changes in shoulder conformation. Some breeds developed breed age average correction factors to standardize for age differences. It must also be remembered that a truly great cow will look good at all stages and ages.

Descriptive, Functional Type Classification

The descriptive type traits were implemented as a part of the classification program by the Holstein-Friesian Association on January 1, 1967. They were subsequently adopted in part or entirely by the other breed associations. The purpose of the descriptive functional traits was to point out strengths and weaknesses in an individual rather than to evaluate them. The description for any particular trait or characteristic was provided by coded numbers that offered approximately five choices for a particular trait. Descriptive terms for both strengths and weaknesses were included. For example, four mammary traits were coded under five headings with five choices for a trait as follows:

Fore Udder:
1. Moderate length and firmly attached.
2. Moderate length and slightly bulgy.
3. Short.
4. Bulgy or loose.
5. Broken and (or) faulty.

Rear Udder:
1. Firmly attached, high and wide.
2. Intermediate in height and width.
3. Low.
4. Narrow and pinched.
5. Loosely attached and (or) broken.

Udder Support and Floor:
1. Strong suspensory ligament.
2. Lack of defined halving.
3. Floor too low.
4. Tilted.
5. Broken suspensory ligament and (or) weak floor.

Teat Size and Placement:
1. Plumb desirable length and size and squarely placed.
2. Acceptable with no serious fault.
3. Rear teats back too far.
4. Wide front teats.
5. Undesirable shape.

The coded numbers for a trait were recorded for stature, head, front end, back, rump, hind legs, feet, fore udder, rear udder, udder support and floor, and teats.

The descriptive trait program has been valuable for dairy farmers and(or) breeders who wish to make a top selection for a service sire or sires from those available at the AI studs. The coded numbers were summarized on a computer in order to provide information on a progeny of a sire. A sire could then be labeled as a herd or breed improver, as an average bull, or as one on the demolition force that destroys rather than builds. Sires were found at all levels and identified with many combinations of strengths and weaknesses, but the precise breeder studied the Sire Performance Summaries until he or she found a sire having a high predicted difference for production and type. Now an index is used to help with this.

The extensive participation of dairy farmers in the functional type trait classification program demonstrated that they appreciated the practical application of the method. The program was geared to recognizing the importance of production and type. It is noteworthy that for the Holstein breed for which the senior author had the responsibility for the revision of the classification system and the introduction of the functional descriptive type traits, there was an increase in the total number of cows classified for 16 succeeding years following the revision, with a final total of 466,600 in a year.

Richard Mansfield in the Centennial Edition, *Progress of the Breed— The History of U. S. Holsteins* (edited by Robert Hastings) commented on the descriptive, functional type traits classification program as follows:

> The new program proved to be popular and would not be replaced until 1983. Since its introduction more than four million registered Holsteins have been evaluated in the United States as well as thousands of others overseas. Classification for type became a major breeding tool and the Association, with a staff of 35 classifiers in the U. S. to evaluate herds at home and abroad, classified five times as many cows at the end of the 16 year life of Descriptive Classification than when it began.

> Not only was the program popular, it also reemphasized the importance of practical, functional type. Breeders began to give greater attention to conformation than at any time before, emphasizing wearability—the ability to produce and reproduce consistently and efficiently for many lactations. While the final score was still important to many, other breeders sought help in pinpointing strong and weak points in animals and selecting sires to improve specific trait weaknesses. Descriptive classification gave them the tools to do it.

UNIFORM TYPE TRAITS APPRAISAL PROGRAM

The classification systems discussed earlier served the purpose of rating animals (excellent, very good, and so on) and assigning them point scores based on the 100-point ideal. The descriptive system did an excellent job of indicating strengths and weaknesses of individual cows and the daughters of bulls. The information was useful for merchandising dairy animals, identifying weaknesses and strengths in individual animals and in groups of animals, and in developing tools (for example, the PDT) which could be used for genetic purposes. However, in summarizing the data gathered from these classifications to guide breeding programs, it was discovered that the results could be advanced by making them more precise for genetic improvement by selection.

The need for a system designed for sire evaluation and one that could be used as an accurate genetic tool with more precise means of trait evaluation rather than a method of describing individual strengths and weaknesses was clear. In view of this, the National Association of Artificial Breeders (NAAB) appointed an ad hoc type committee in 1977. This committee was charged with reviewing the current type evaluation systems and designing a new program that would provide the type information needed to increase the rate of genetic progress in dairy cattle conformation. A recommendation was made to work with all of the dairy breed associations for implementation of a system on an industrywide basis. Below is a summary of the basic concepts of the recommendations adopted by the NAAB Board of Directors.

The committee decided to design a new type evaluation program and develop seven basic concepts:

1. Evaluate single biological traits which can significantly increase the potential for genetic improvement and do so independently of other traits.
2. Evaluate traits by using numerical scores from one biological extreme to the other (linear scoring). This linear scoring would not describe a trait as compared to an ideal; instead, it would measure the degree of that trait. A linearized system could be applied uniformly for all traits of economic and functional importance. For example, rear legs (side view) would be scored from most sickled to straightest, even though the most desirable would not be at either end of the scale. Also, a numerical range wide enough to establish a high degree of accuracy would be used. A range of 50 points was recommended.
3. Score all similar-aged contemporaries so that the data could be analyzed on a within-contemporary group.
4. Evaluate both registered and grade cows so that larger numbers could be scored to secure more data more quickly for sire summaries.
5. Evaluate animals while they are young in order to obtain information to summarize sires as soon as possible.
6. Score the cows without previous knowledge of the sire or previous score in order to remove bias.
7. Analyze the data by using the best linear unbiased predictor (BLUP) procedure, because this is the most accurate method known to evaluate the data.

The next step in developing the system was to decide which traits should be evaluated. The objectives were to do the following:

1. Avoid historical biases when reviewing traits.
2. Make sure each trait was of economic value, clearly understood, and defined such that only one trait was measured at a time.
3. Finalize a list of primary traits, the number of which could not exceed what was reasonable for field application, remembering that this system describes degree, not models.

Based on these criteria, 13 traits were selected to be included in the score card: stature, chest and body (considering age and stage of lactation), dairy character (independent of performance), foot shape (angle), rear legs (side view), pelvic angle, rump width, fore udder attachment, rear udder width (at attachment), rear udder height (at attachment), teat placement (rear view), suspensory ligament (cleft), and udder depth (relative to point of hock). The score card also included areas for remarks and defects (see Fig. 413).

NAME OF ANIMAL		REGISTRATION NO.		DATE OF BIRTH

TATTOO	CHAIN NO.	MOST RECENT CALVING	BARN NAME	*

Stature	Chest and Body	Dairy Character	Foot Shape (angle)	Rear Legs (side view)	Pelvic Angle	Rump Width	Fore Udder Attachment	Rear Udder Width	Rear Udder Height	Teat Placement (rear view)	Suspensory Ligament (cleft)	Udder Depth	Final Score	Remarks	Defects

Remarks

10 Dry
11 Stale
12 Abnormal calving
13 Blemished udder
14 Congested udder
15 Out of condition

Defects

20 Frail
21 Compact
22 Low front
23 Open shoulder
24 Weak jaw
25 Eye defect
26 Weak loin
27 Coarse

Defects (cont.)

30 Front feet toe out
31 Close at hocks
32 Shallow heel
33 Spread toes
34 Curled toe
35 Weak pasterns
36 Crampy
37 Coarse hocks

40 Recessed anus
41 High or coarse tailhead
42 Thurls too far back
43 Narrow pins

Defects (cont.)

50 Short fore udder
51 Excessively long fore udder
52 Lacks fore udder capacity
53 Udder tilt—low rear
54 Udder tilt—low front
55 Undesirable udder quality
56 Chronic edema

60 Rear teats back too far
61 Teats close on side
62 Pencil teats
63 Teats too large
64 Teats too small
65 Strutting teats

Figure 413   Uniform Functional Type Appraisal Report (Courtesy W. D. Hoard and Sons Co., Ft. Atkinson, WI)

1. Stature (measured at withers)

90-99 Very tall   80-89 Tall   70-79 Intermediate   60-69 Low set   50-59 Very low set

2. Chest and Body (considering age and stage of lactation)

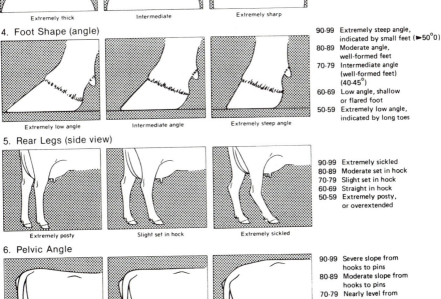

90-99 Wide chest, deep rib, long body
80-89 Wide and strong, moderate length of body
70-79 Intermediate
60-69 Narrow
50-59 Extremely narrow and frail

Extremely narrow and frail   Intermediate   Wide chest, deep rib

3. Dairy Character (independent of performance)

90-99 Extremely sharp
80-89 Clean cut, open and angular
70-79 Intermediate
60-69 Thick and coarse
50-59 Extremely thick and tight-ribbed

Extremely thick   Intermediate   Extremely sharp

4. Foot Shape (angle)

90-99 Extremely steep angle, indicated by small feet (►50°0)
80-89 Moderate angle, well-formed feet
70-79 Intermediate angle (well-formed feet) (40-45°)
60-69 Low angle, shallow or flared foot
50-59 Extremely low angle, indicated by long toes

Extremely low angle   Intermediate angle   Extremely steep angle

5. Rear Legs (side view)

90-99 Extremely sickled
80-89 Moderate set in hock
70-79 Slight set in hock
60-69 Straight in hock
50-59 Extremely posty, or overextended

Extremely posty   Slight set in hock   Extremely sickled

6. Pelvic Angle

90-99 Severe slope from hooks to pins
80-89 Moderate slope from hooks to pins
70-79 Nearly level from hooks to pins
60-69 Pins higher than hooks
50-59 Pins clearly higher than hooks

Pins higher than hooks   Nearly level   Severe slope

7. Rump Width

90-99 Extreme width of pelvic area
80-89 Wide width of pelvic area
70-79 Intermediate width of pelvic area
60-69 Narrow in pelvic area
50-59 Extremely narrow in pelvic area

Extremely narrow   Intermediate width   Extreme width

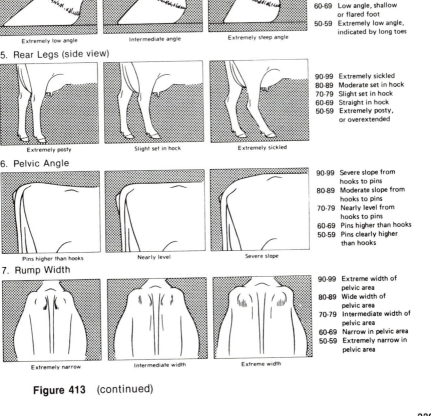

Figure 413 (continued)

8. Fore Udder Attachment

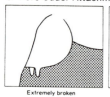

Extremely broken   Intermediate strength   Extremely tight attachment

90-99 Extremely tight
       attachment
80-89 Strong, firm
70-79 Intermediate strength
60-69 Loose attachment
50-59 Extremely broken

9. Rear Udder Width (at attachment)

Extremely narrow   Intermediate width   Extremely wide

90-99 Extremely wide
80-89 Wide
70-79 Intermediate width
60-69 Narrow
50-59 Extremely narrow

10. Rear Udder Height (at attachment)

Extremely low   Intermediate   Extremely high

90-99 Extremely high
80-89 High
70-79 Intermediate
60-69 Low
50-59 Extremely low

11. Teat Placement (rear view)

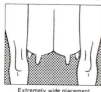

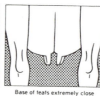

Extremely wide placement   Centrally placed   Base of teats extremely close

90-99 Base of teats
       extremely close
80-89 Teats too close
70-79 Centrally placed, nearly
       aligned, tilting inward
60-69 Intermediate placement,
       front teats wider than
       rear teats
50-59 Extremely wide
       placement

12. Suspensory Ligament (cleft)

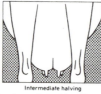
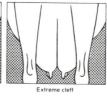

Broken   Intermediate halving   Extreme cleft

90-99 Extreme cleft
80-89 Clearly defined halving
70-79 Intermediate halving
60-69 Flat udder floor
50-59 Broken

13. Udder Depth (relative to point of hock)

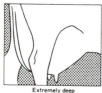

Extremely deep   Udder floor at hock   Extremely shallow

90-99 Extremely shallow,
       udder floor well
       above hock
80-89 Udder floor clearly
       above hock
70-79 Udder floor at
       the hock
60-69 Udder floor clearly
       below hock
50-59 Extremely deep

Figure 413 (continued)

These concepts served as the foundation for the various breed associations to develop and implement a linear method of rating type traits of economic and functional importance to each breed. Thus, the traits that are evaluated differ among the various breed linear type programs. In addition to evaluating type traits, some programs include management traits, such as temperament, milking speed, and so on. This is an attempt to evaluate non-type traits that can affect the profitability of a herd.

The type data will be summarized comparing daughters of sires to their herd mates in such a way to account for genetic merit of contemporary herd mates, to adjust for genetic trends, and to rank all bulls relative to a given point in time using the BLUP procedure.

The primary purpose of this classification program, as compared to former programs, is to provide more precise data for selection and genetic improvement. The system may have to be revised over time because experience has shown the need for including more or fewer traits or other modifications. Data should also more accurately determine the contribution of each of the traits in determining herd life and lifetime production.

Using this system need not negate the merchandising value of classification. A column for final score (as indicated in Fig. 413) can be included in the score sheet. If a breed desires breakdown scores for general appearance, dairy character, body capacity, and mammary system, these scores could also be included. The NAAB committee recommended that this be left to the discretion of the various breed associations. Figure 413 is a copy of the Uniform Type Appraisal Report.

All breeds have adopted the new linear type classification system, but several breeds use it with slight modifications.

Type summaries based on data from the new system have been published. An example of the Jersey Sire Type PDs for appraisal traits is presented on page 342.

The first four columns of the summary give the production type index (PTI), its repeatability, the number of daughters evaluated, and the resulting type repeatability. The next column is the PD of final score followed by the PDs of the uniform type appraisal traits. The values for final score and the 13 traits of the individual bull (Sample Jersey Bull) are expressed relative to the Jersey breed average, which is listed just below the bull's PDs for comparison purposes. Columns 5 through 8, 12 though 15, and 17 are expressed as plus (+) or minus (−) from breed average. For example, Sample Jersey Bull is + 0.8 for stature. This means that if a large number of breed average Jersey cows were bred to Sample Jersey Bull, we would expect his daughters to average 75.9 in stature (breed average 75.1 + PD 0.8 = 75.9).

The other five categories (9, 10, 11, 16, and 18) include both an alphabetic and a numerical value. In these categories the numerical value represents the PD, which is indicative of the amount of change a breeder can expect; the alphabetic character varies for different traits, but it still indicates which way from breed average the change will occur. The meaning of the alphabetic characters for the five traits is:

Foot shape (angle)	L	=	low angle, S = steep angle
Rear legs (side view)	P	=	posty, S = sickled
Pelvic angle	H	=	pins high, L = pins low
Teat placement (rear view)	W	=	wide, C = close
Udder depth (relative to point of hock)	D	=	deep, S = shallow

For example, Sample Jersey Bull had a value of SO.5 for udder depth (column 18). This means that if a large number of average Jersey cows were bred to

Predicted Differences for Appraisal Traits[a]

	Prod Type Index	Prod Type Rep	No Dau Class	Type Rep	Final Score	Stat
Sample Jersy Bull	+112	41%	20	50%	0.1	0.8
Breed Average					77.0	75.1
Column	(1)	(2)	(3)	(4)	(5)	(6)
	Chest And Body	Dairy Char	Foot Angle Low Steep	Rear Legs Posty-Sickle	Pelvic Angle Pins Hi-Lo	Rump Wid
Sample Jersey Bull	1.5	0.3	SO.7	PO.1	LO.3	−0.3
Breed Average	81.4	82.8	68.5	80.7	78.7	77.3
Column	(7)	(8)	(9)	(10)	(11)	(12)
	Fore Udd Att	Rear Udd Wid	Rear Udd Hgt	Teat Place Wide-Close	Susp Lig	Udd Depth Deep-Shal
Sample Jersey Bull	0.9	−0.1	−0.4	WO.5	0.2	SO.5
Breed Average	79.8	78.4	82.3	68.2	80.2	80.9
Column	(13)	(14)	(15)	(16)	(17)	(18)

[a]From the *Jersey Journal* (February 1981), 19.

Sample, we would expect his daughters to average 0.5 shallower or 81.4 in udder depth (breed average 80.9 + PD 0.5 = 81.4).

The table on this page lists nine Jersey bulls ranked by PTI and includes the same information as on Sample Jersey Bull. The recommended method of using this information to make genetic progress for both type and production

Predicted Differences for Appraisal Traits[a]

Bull	Prod Type Index	Prod Type Rept	Num Dau Appr	Type Rept	Final Score	Stat
A	348	50	19	48	2.2	0.1
B	308	98	789	97	0.6	1.8
C	279	58	26	54	2.4	3.0
D	279	62	49	62	-0.2	-0.4
E	277	48	37	58	2.1	1.2
F	272	57	34	58	0.9	-1.4
G	261	56	41	55	0.2	0.4
H	254	81	78	74	-0.2	-0.3
I	249	64	36	62	0.4	-2.7

Bull	Chest And Body	Dairy Char	Foot Angle Low-Steep	Rear Legs Posty-Sickle	Pelvic Angle Pins Hi-Low	Rump Wid
A	0.4	1.0	LO.3	PO.6	LO.1	1.0
B	0.3	2.4	L1.3	S1.7	LO.1	1.9
C	1.1	0.2	SO.5	0.0	H1.2	2.7
D	-0.6	1.1	SO.5	PO.8	LO.2	-0.9
E	0.3	1.4	LO.3	SO.1	HO.1	0.2
F	-0.3	0.5	LO.4	SO.9	HO.1	0.1
G	-0.2	0.8	LO.6	SO.1	L1.0	0.0
H	-0.2	0.7	LO.3	SO.9	LO.2	0.3
I	-2.1	0.5	LO.5	PO.6	LO.3	-2.6

Bull	Fore Uddr Att	Rear Uddr Wid	Rear Uddr Hght	Teat Place Wide-Close	Susp Lig	Udder Depth Deep-Shalo
A	0.3	0.5	1.1	CO.9	0.4	DO.2
B	-1.0	1.2	0.8	WO.5	0.4	CO.6
C	1.9	1.6	1.4	CO.9	-0.4	SO.9
D	-0.9	-2.0	-0.4	WO.2	-0.2	DO.3
E	1.9	1.4	1.2	CO.4	1.8	SO.7
F	0.4	0.7	0.7	WO.4	-0.4	DO.4
G	-0.7	-0.3	-0.8	CO.5	-0.2	D1.0
H	-0.7	-0.4	0.4	WO.7	0.1	0.0
I	0.1	-1.6	0.1	CO.9	1.5	0.0

[a]From the *Jersey Journal* (May 1981), 35.

is for the breeder to first select bulls for use in the herd on the basis of their ability to improve production (PD milk, fat, and so on) or PTI. Then, using the unbiased evaluation of each cow in the herd or sire summary information of sires of groups of daughters in the herd, the breeder chooses those bulls whose type summary information indicates they are improvers for those minus traits (−); that is, the breeder uses corrective mating. The U.S. Department of Agriculture sire lists conveniently start with the highest production or production type index bulls first. When selecting bulls to use, it is important that the breeder start at the top of this list and work down—and not too far down lest the breeder greatly reduce genetic progress for production. It is also recommended that the breeder limit corrective mating to one or two traits per mating so as not to dilute progress for any one trait.

It is the authors' opinion that the Uniform Type Traits Appraisal program offers an exciting opportunity to hasten the genetic progress for type improvement, similar to the progress made by widespread use of PD for production which started in the late 1960s. Simply stated, it should add more science to the art and science of breeding superior dairy cattle.

MULTIPLE EXCELLENTS

Multiple E classification, earned in different age brackets and designated 2E, 3E, 4E, and 5E, was started by the Brown Swiss breed and taken up by most other breeds. The Holstein association included the multiple E designation when the switch was made to the functional type traits. It denotes that a cow improves with age or maintains a high standard once it is reached.

This has been a popular program and labels the good ones. To demonstrate how it works in practice, a 4E 95 All-American Ayrshire cow with 25,081 lb of milk is presented in Figures 414 to 417.

RECORD BREAKERS

Achievements reached in all phases of production verify the fact that great progress has been made in the genetic improvement of dairy cattle, together with having management procedures and the type of dairy cows to provide the kind of machine to take full advantage of this.

A good example of this is the top records made in 305 and 365 days and also the lifetime total production. Furthermore, type and production are now combined far beyond previous experience. An example of this is the All-Time All-American cow, with 48,731 pounds of milk, 4.2%, and 2028 pounds of fat at 7 years-7 months in 365 days. She is the highest milk producer among daughters of a most outstanding bull with 54,843 tested daughters (+366 lb milk) and 40,961 classified daughters at an average of 83 points, plus 2.19 PDT. This bull has 2862 Excellent daughters and 18 daughters that were voted All-Americans.

Figure 414   1 Ex 91.2 at 4 yrs, All-American three-year-old, Grand NYSF.
3y-10 months, 305 days   15,714 1 lb milk, 3.9%   615 lb fat.
(Courtesy Charles and Kenneth Burr, Burr-Ayr Farms, Trumansburg, NY)

Figure 415   2 Ex 90.8 at 6 years.
5y-11 months, 305 days   17,500 lb milk, 3.7%   651 lb fat.
(Courtesy Charles and Kenneth Burr, Burr-Ayr Farms, Trumansburg, NY)

Figure 416 3 Ex 94 at 9 years.
9y-1 months, 305 days   22,721 lb. milk, 3.2%   724 lb fat.
(Courtesy Leon A. Perry, Stone House Farm, Orleans, VT)

Figure 417 4 Ex 95 at 11 years. Good producer adding to her lifetime total
of 106,322 lb milk, 3.7%, 3875 lb fat.
(Courtesy Leon A. Perry, Stone House Farm, Orleans, VT)

Figure 418 This 4E Across Brown Swiss Cow has many achievements and is very worthy of the Multiple Excellents. Her dam was a 5E cow with good production at an advanced age. The Brown Swiss breed scores cows up to Excellent, but does not provide a score for cows that score 90 or above. The cow pictured here was nominated All-American seven times. She was voted All-American 4-year-old and All-American Aged Cow and twice was a member of the All-American Get of Sire. She was the National Performance Winner and a National Protein Award Winner. She is on the "Elite" list for production and some of her records follow:

5y-10m	365d	2X	23,010 lb milk	4.5%	1026 lb fat
6y-11m	365d	2X	25,220 lb milk	4.2%	1053 lb fat
8y-5 m	365d	2X	27,530 lb milk	4.0%	1110 lb fat
10y-1 m	365d	2X	30,210 lb milk	4.1%	1224 lb fat

(3.8% protein 1158 lb)

This cow frequently milked over 100 lb a day and she is a multiple E cow with a good combination of functional type, continuous high production, high classification scores, and twice All-American. (Courtesy Top Acres, Wayne E. Slyker and family, St. Paris, OH)

It is very impressive that the last two pictures in this book show two cows with the world's record in both the 305 and 365 day on 2X milking, both made within a year, one in Pennsylvania and the other in Indiana. The first peaked at 180 lb of milk and averaged 139 lb of milk a day for the entire year. Not to be outdone, the Indiana cow peaked at 195 lb of milk per day and averaged 152 lb for the entire year on 2X milking. This is phenomenal. Both cows had a good machine with a respective classification of Excellent 92 and 91.

Also noteworthy is the high lifetime production figure achieved with the top record holder Breezewood Patsy Bar Pontiac, a 6E 93 GMD Holstein cow, in Ohio. She accumulated a total of 425,769 lb milk, 4.5%, 19,203 lb fat in 5267 days at 18 years, 6 months of age. Her highest record was at 10 years, 10 months, 365 d, 2X, 47,500 lb milk, 4.7%, 2230 lb fat.

It is a foregone conclusion that all these cows with a record high production always have a big appetite. When this was experimentally tested at Cornell and other universities in the United States and Europe, it was found that appetite had a high heritability. However, it also had a very high association (correlation) with high milk production and, therefore, it was not necessary to test for appetite, because testing and selection on milk production was sufficient.

CONCLUDING REMARKS

In the final analysis, both the highly specialized dairy cattle breeder and the practical dairy farmer, including the strictly commercial producer, can justify considerable emphasis on type, provided that high production accompanies it.

The interplay between production and functional type in an operation and selection program is now clearly understood and is no longer controversial.

Type characteristics have considerable influence on correct management practices. A good example is the discrimination against fat in heifers, which later impairs udder quality and production.

The emphasis on type is not an academic development at all, but is demanded by the breeders of good cattle. Good type or eye appeal, especially dairy

Figure 419 Outstanding Holstein cow at almost 10 years of age. She has a classification of Ex 96 and produced 31,028 lb of milk and 1020 lb of fat at 5 years of age. She is the kind that also excels in the show ring. Her many winnings included All-American Senior Yearling and Reserve All-American 2- and 4-year-old. This cow was born and raised on a practical dairy farm. With over 17,000 lb of milk in her first lactation, when she won the Reserve All-American, she is the ultimate for production, high classification score, and show-ring winnings. (Courtesy Collins-Crest Farm, Perry, NY)

character with strength, a well-supported udder, and good feet and legs help make sales of surplus females. Owners, or their hired representatives, who appreciate good type get more satisfaction and enjoyment out of their cattle, and this is often reflected in better care and the resulting higher economic returns.

Finally, participating in 4-H, FFA, and other agricultural vocational youth programs that have projects involving good dairy cattle has helped many young people to do a better job and to become successful. It is a fact that it is the best known means to inspire youth and show them a way of life which they can attain by their own self-disciplined, inspired effort. The products of human accomplishments derived therefrom show such encouraging positive progress that there is a new spirit of confidence in this type of activity. The breeders of good cattle should be complimented for their excellent cooperation in working with youth and assuming responsibility to assure development.

Figure 420   There is a real satisfaction in planning a mating and then watching a calf go through the heifer stage and develop into a high producer, a good show cow, or an efficient reproducer. The object is to produce offspring that make good in a practical, highly developed commercial herd. This cow had the world's record for milk in 305 and 365 days. She is the ultimate in the chain of events of special interest to a dairy farmer. At 9 years, 8 months of age on 2X milking she produced more milk than any other cow ever did before in the entire history of dairying. With a classification of twice Excellent and a score of 92 she has the machine required for her tremendous accomplishment. In 305 days she produced 44,144 lb. of milk, 3.1%, and 1360 lb of fat. In 365 days she produced 50,759 lb of milk, 3.04%, and 1548 lb of fat. She reached a peak of 180 lb of milk and averaged 139 lb of milk a day for the entire year. She exceeded the previous world's record by over 5000 lb. (Courtesy Clarence and Kenneth Mowry, Roaring Spring, PA)

Figure 421 Another world's milk record. This last picture displays a cow, Beecher Arlinda Ellen (Ex 91), that established a new phenomenal world's milk record a year after the record was established by the cow in Fig. 420. Arlinda Ellen made a new record for both 305 and 365 days with 50,314 and 55,661 lb of milk, respectively. This was on 2X milking at 5 years, 9 months of age. She accomplished this at a peak of 195 lb of milk per day and an average production of 152 lb for the entire year. The cow in Fig. 420 and the cow here establish a type pattern. Both possess an abundance of strength with dairy quality. Arlinda Ellen was bred for production. Her sire was Pawnee Farm Arlinda Chief, a proven plus bull with a high predicted difference for milk in his daughters. He was used extensively as a herd and breed improver in the AI program. (Courtesy Harold L. Beecher family, Rochester, IN)

INDEX